electronics one

Hayden Electronics One-Seven Series

Harry Mileaf, Editor-in-Chief

electronics one — Electronic Signals □ Amplitude Modulation □ Frequency Modulation □ Phase Modulation □ Pulse Modulation □ Side-Band Modulation □ Multiplexing □ Television Signals □ Navigation Signals □ Facsimile □ Heterodyning □ Harmonics □ Waveshaping

electronics two — Electronic Building Blocks □ Basic Stages □ The Power Supply □ Transmitters □ Receivers □ UHF □ Telemetry □ Television □ Radar and Sonar □ RDF □ Radio Navigation □ Radio Control □ Quad Sound

electronics three — Electron Tubes □ Diodes □ Triodes □ Tetrodes □ Pentodes □ Multielement Tubes □ Gas-Filled Tubes □ Phototubes □ Electron-Ray Indicators □ Cathode-Ray Tubes □ UHF and Microwave Tubes □ Magnetrons □ Klystrons □ The Traveling-Wave Tube

electronics four — Semiconductors □ P-N Diodes □ Avalanche Diodes □ Switching Diodes □ Zener Diodes □ Tunnel Diodes □ N-P-N and P-N-P Transistors □ Tetrode Transistors □ Field Effect, Surface Barrier, Unijunction, SCR Transistors □ Photodevices

electronics five — Power Supplies □ Rectifiers □ Filters □ Voltage Multipliers □ Regulation □ Amplifier Circuits □ A-F, R-F, and I-F Amplifiers □ Video Amplifiers □ Phase Splitters □ Follower Amplifiers □ Push-Pull Amplifiers □ Limiters

electronics six — Oscillators □ Sinusoidal and Nonsinusoidal Oscillators □ Relaxation Oscillators □ Magnetron Oscillators □ Klystron Oscillators □ Crystal Oscillators □ Modulators □ Mixers and Converters □ Detectors and Demodulators □ Discriminators

electronics seven — Auxiliary Circuits □ AVC, AGC, and AFC Circuits □ Limiter and Clamping Circuits □ Separator, Counter, and Gating Circuits □ Time Delay Circuits □ Radio Transmission □ Antennas □ Radiation Patterns □ R-F Transmission Lines

electronics one

HARRY MILEAF EDITOR-IN-CHIEF

HAYDEN BOOK COMPANY, INC.
Rochelle Park, New Jersey

ISBN 0-8104-5954-X
Library of Congress Catalog Card Number 75-45505

Printed in the United States of America

3	4	5	6	7	8	9	PRINTING
78	79	80	81	82	83	84	YEAR

preface

This volume is one of a series designed specifically to teach electronics. The series is logically organized to fit the learning process. Each volume covers a given area of knowledge, which in itself is complete, but also prepares the student for the ensuing volumes. Within each volume, the topics are taught in incremental steps and each topic treatment prepares the student for the next topic. Only *one* discrete topic or concept is examined on a page, and *each* page carries an illustration that graphically depicts the topic being covered. As a result of this treatment, neither the text nor the illustrations are relied on solely as a teaching medium for any given topic. Both are given for *every* topic, so that the illustrations not only complement but reinforce the text. In addition, to further aid the student in retaining what he has learned, the important points are summarized in text form on the illustration. This unique treatment allows the book to be used as a convenient review text. Color is used not for decorative purposes, but to accent important points and make the illustrations meaningful.

In keeping with good teaching practice, all technical terms are defined at their point of introduction so that the student can proceed with confidence. And, to facilitate matters for both the student and the teacher, key words for each topic are made conspicuous by the use of italics. Major points covered in prior topics are often reiterated in later topics for purposes of retention. This allows not only the smooth transition from topic to topic, but the reinforcement of prior knowledge just before the declining point of one's memory curve. At the end of each group of topics comprising a lesson, a summary of the facts is given, together with an appropriate set of review questions, so that the student himself can determine how well he is learning as he proceeds through the book.

Much of the credit for the development of this series belongs to various members of the excellent team of authors, editors, and technical consultants assembled by the publisher. Special acknowledgment of the contributions of the following individuals is most appropriate: Frank T. Egan, Jack Greenfield, and Warren W. Yates, principal contributors; Peter J. Zurita, S. William Cook, Jr., Steven Barbash, Solomon Flam, and A. Victor Schwarz, of the publisher's staff; Paul J. Barotta, Director of the Union Technical Institute; Albert J. Marcarelli, Technical Director of the Connecticut School of Electronics; Howard Bierman, Editor of *Electronic Design;* E. E. Grazda, Editorial Director of *Electronic Design;* and Irving Lopatin, Editorial Director of the Hayden Book Companies.

HARRY MILEAF
Editor-in-Chief

contents

what is electronics?

Now that you are beginning the study of electronics, it might be a good idea to ask yourself, "Do I really know what electronics is?" From your familiarity with some of the everyday applications of electronics, you know that there is a relationship between *electronics* and *electricity*, since electronic devices, such as radios and televisions, all use electricity. However, all devices that use electricity are not electronic devices. The washing machine, electric iron, television, and hi-fi all use electricity; but the washing machine and the iron are *electrical* devices, while the television and hi-fi are *electronic* devices. What makes the difference? The answer can be found in the concept of *intelligence*, and it is on this basis that we will define the study of electronics.

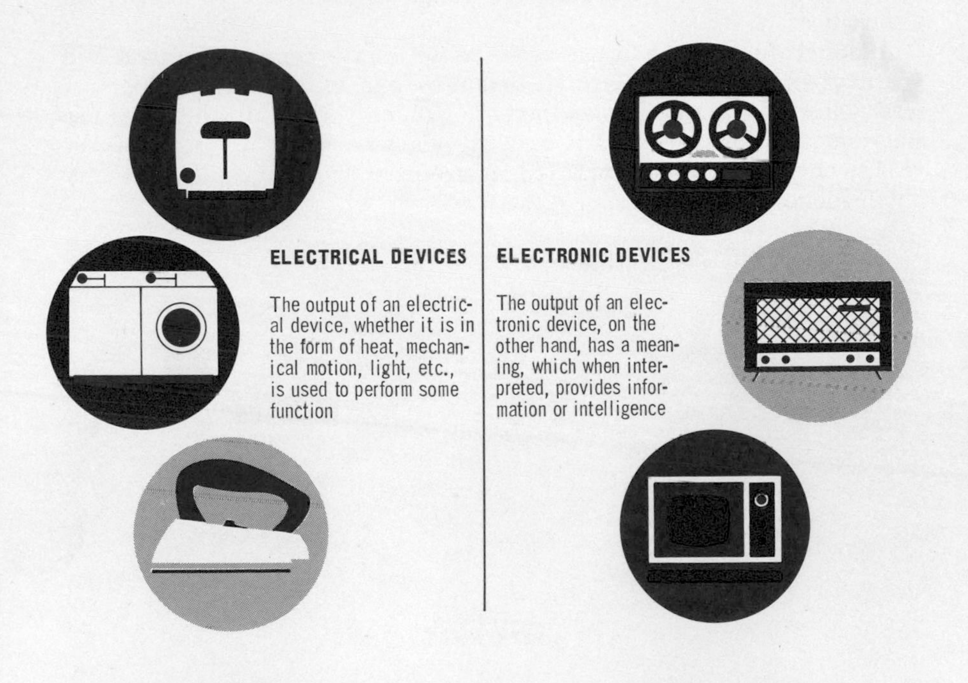

In the study of *electricity*, you learned how electrical phenomena are used to provide *power* or *energy*. Thus, an electric iron is an electrical device because electricity is used to provide energy, in the form of heat, by its element. Similarly, electricity provides the power to turn the tub of a washing machine as well as to operate its control relays. In the study of *electronics*, though, you will learn how electricity is used to carry *intelligence*. You will find that this intelligence varies widely, from the simple doorbell, which tells you that someone is calling, to complex radar systems, which locate and track fast-moving distant targets. Nevertheless, any device that uses electricity to *tell*, *show*, or otherwise *inform* is electronic.

where is electronics used?

There is probably no aspect of our present-day life that has not been influenced to some extent by electronics. Electronic devices and equipment are used in so many ways, and for so many reasons, that it is almost impossible to adequately summarize their uses.

In certain fields, electronics plays such an important role that it is safe to say that without it, these fields could never have developed to their present status. An example of one of these fields is *communications*. Without electronic radio transmitters and receivers, rapid communication as we know it today would be impossible. Another example is industrial *automation*. Most of the monitoring and control devices that make automation possible are electronic. *Data processing* and *scientific and medical research* are other fields that rely heavily on electronics.

Although not a field in the same sense as are communications and automation, the *military establishment* is one of the largest users of electronics and electronic equipment. Not only does the military use electronics extensively, but it also spends vast sums of money every year to develop new and improved electronic equipment, as well as new applications of electronics.

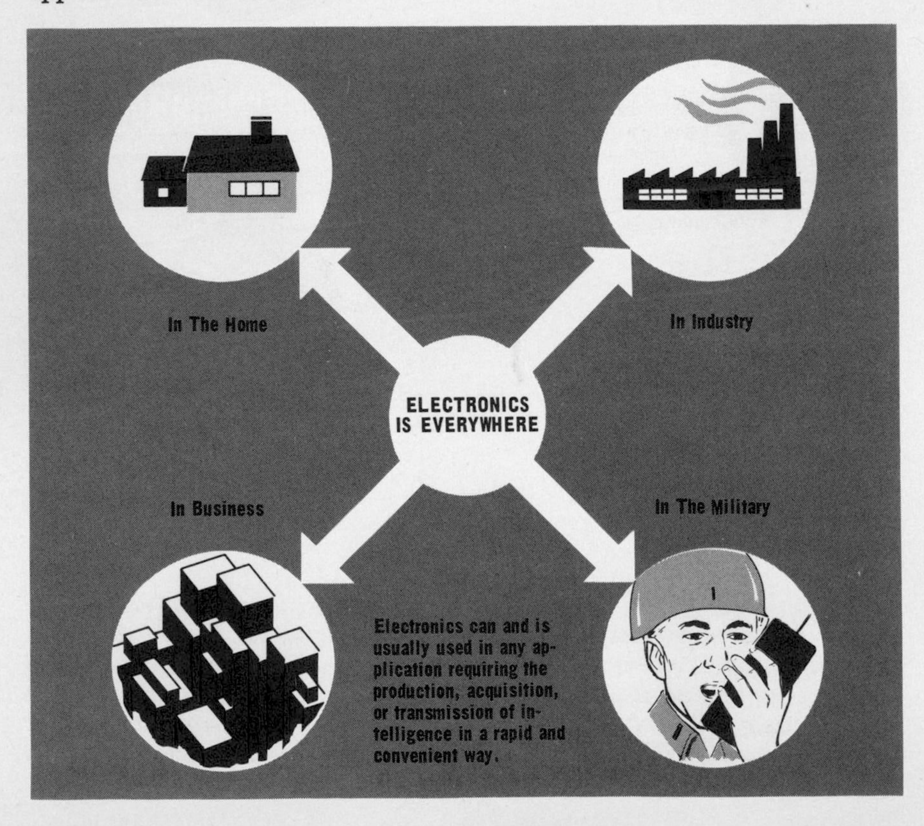

electronic signals

You will recall from what you learned about electricity that electrical currents and voltages have various *characteristics*. These range from the amplitude for d-c currents and voltages to the a-c sinusoidal waveform, frequency, period, etc. Although these characteristics must be considered when you analyze electrical circuits, the quantity you are most interested in is usually the *energy* or *power* delivered to the load.

In electronics, characteristics of signals that are extremely important because of their intelligence-carrying capability include:

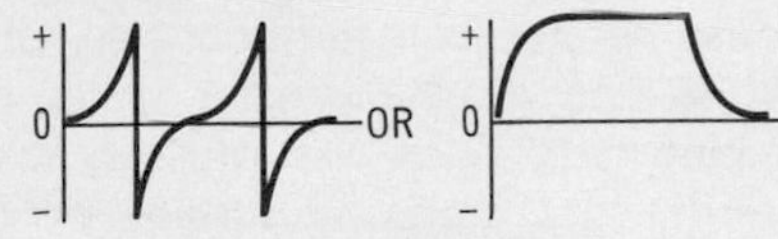

SHAPE

FREQUENCY

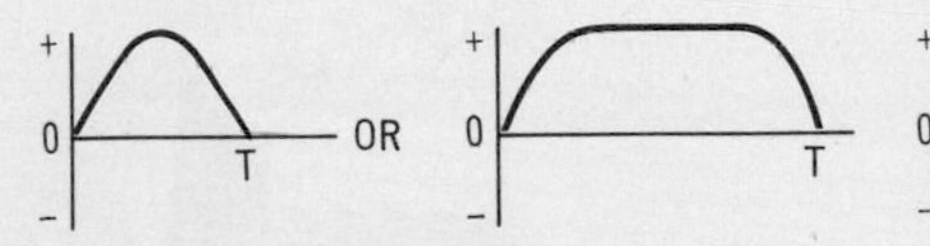

DURATION

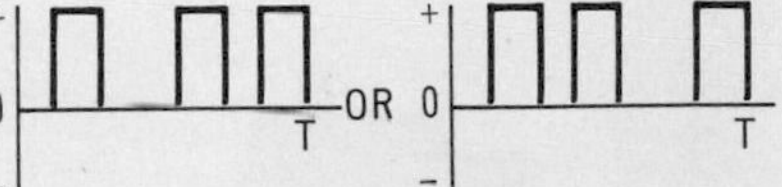

POSITION IN TIME

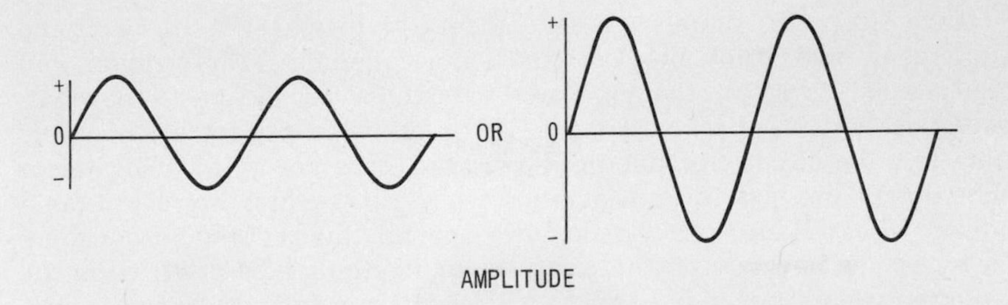

AMPLITUDE

In the study of electronics, the situation is entirely different. The power delivered is still a consideration, but it is no longer the most important characteristic. The other characteristics of the currents and voltages are just as, if not more, important, since it is by means of these characteristics that the currents and voltages *carry intelligence*. For convenience, currents or voltages that carry intelligence are often referred to as *signals*.

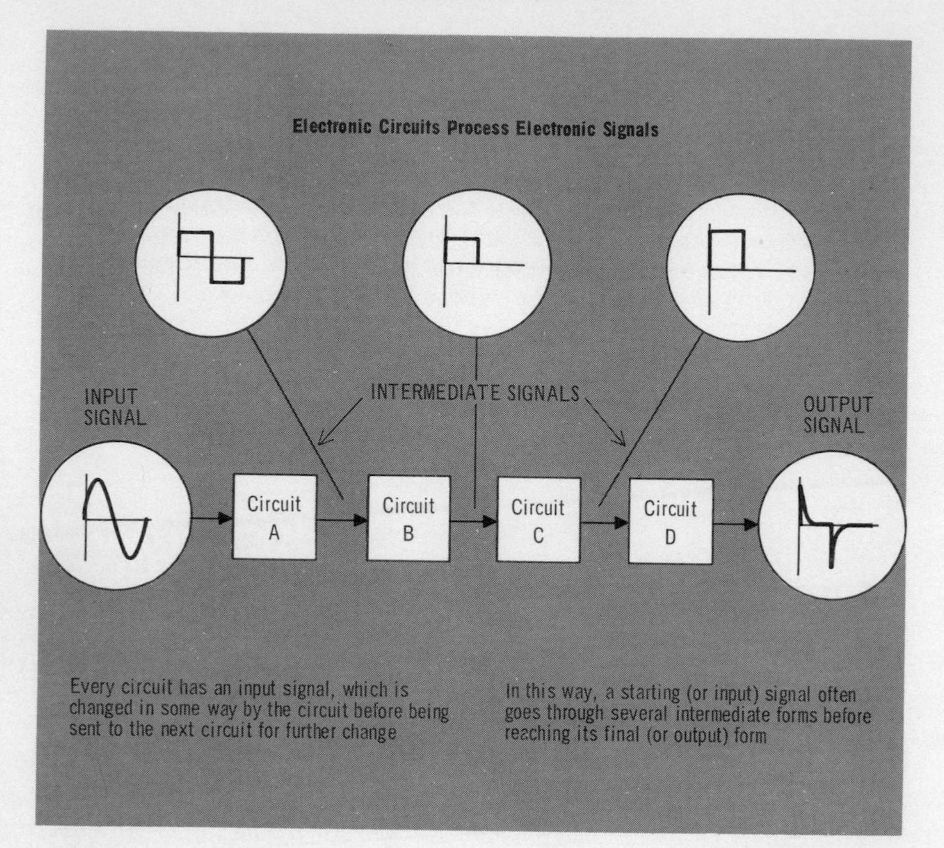

why study electronic signals?

Since electronic signals are the carriers of intelligence in electronic equipment, any study of electronics must include a description and analysis of the most common types of signals. In the past, the characteristics and uses of most types of signals were taught at the *same time* that the circuits producing the signals were described. This was a satisfactory method for a long time. In recent years, however, rapid advances in the electronics field have opened this method to question. New circuits, new equipment, and new applications have resulted in the situation where the same basic type of signal is produced and used in a *wide variety* of ways. It seems desirable, therefore, to *divorce* signals from circuits or equipment as much as possible, and describe them from the standpoint of how they carry intelligence, how they interact with other signals, etc. This somewhat unique approach is used in this volume, which is devoted entirely to electronic signals.

Once you have a firm grasp of the principles and characteristics of electronic signals, you will find that circuits as well as entire equipments can be explained on the basis of them.

d-c signals

The use of dc amply illustrates how intelligence can be added to a voltage or current to create a signal. Consider first the case of two men separated by a great distance. If their only means of communication is a d-c circuit consisting of a battery and a current-limiting resistor, it is obvious that they cannot communicate. A steady d-c current flows in the circuit, and such a current carries no intelligence.

No intelligence can be carried by a constant value of dc

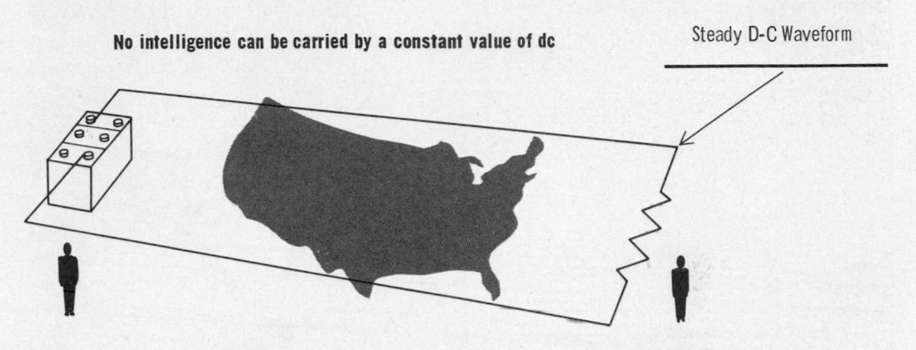

By the addition of a *switch* into the circuit, though, the current can drop to zero when the switch is opened, and rise to its steady value when the switch is closed. So by opening and closing the switch, one man causes the current to either *flow* or *not flow*. The waveform of the current is a pulse when current flows and no pulse when current stops. If the men have a prearranged code whereby each letter of the alphabet is represented by a particular combination of pulses, the man with the switch can then cause the current waveform to carry any message he wants. This, of course, assumes that the other man has a way of "seeing" or "hearing" the waveform.

This example, although impractical, serves to illustrate that intelligence can only be carried when some characteristic of a current or voltage is made to *change* or *vary* in a meaningful way.

If dc is broken up into a series of pulses whose characteristics correspond to some code, the dc then makes up an intelligence-carrying signal

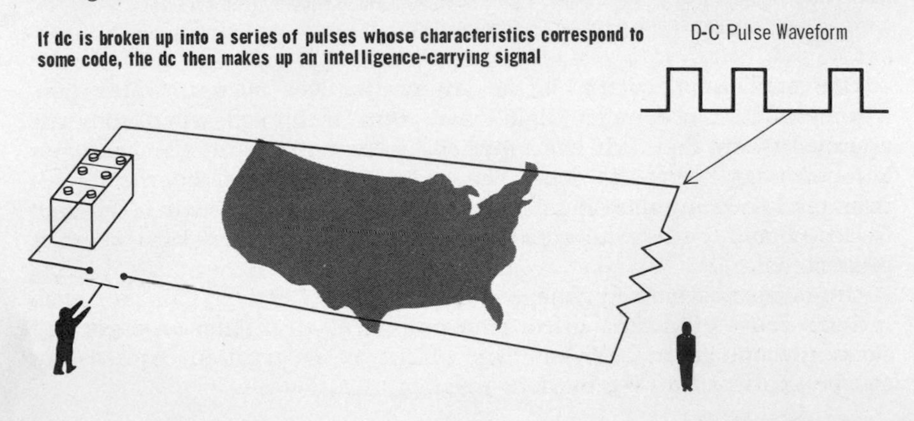

applications

As you will learn, d-c signals are not used as widely as are a-c signals, but they have many practical applications. Some applications make use of *photoelectric cells.* When light strikes a photoelectric cell, the cell emits electrons to produce a d-c current. Thus, when no light strikes the cell, there is no d-c current; and, within certain limits, when light does fall upon the cell, the *amplitude* of the d-c current is proportional to the light intensity. The d-c current, therefore, can serve as a signal which represents the *presence,* or *intensity,* of light striking the cell.

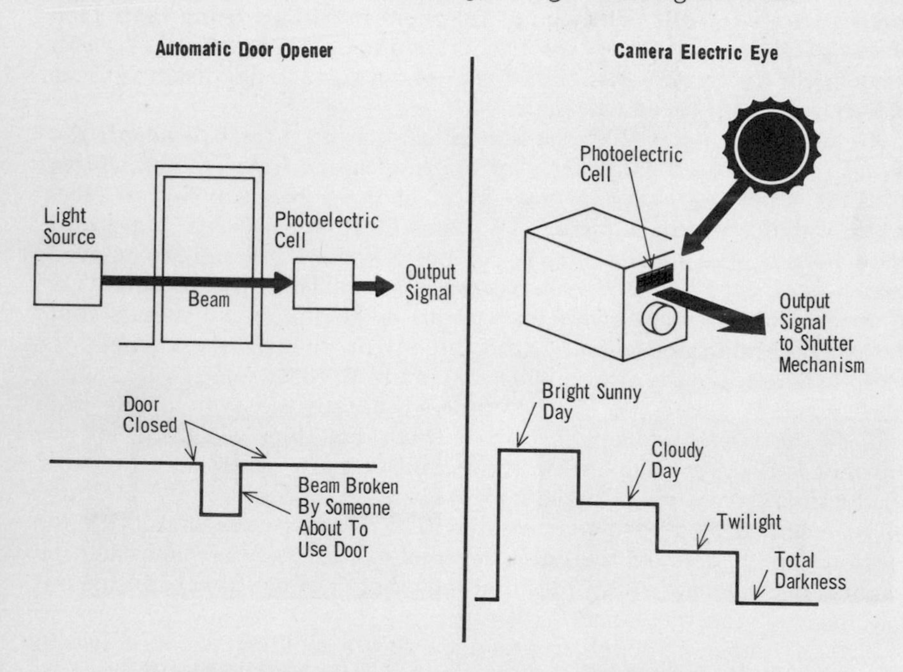

Since dc has only two characteristics that can be varied (presence or absence, and amplitude), all d-c signals have a basic similarity. But the intelligence carried by d-c signals can represent a wide variety of things

One application of this is the automatic door opener, which uses a photoelectric cell and a light beam that is broken when someone approaches the door. The interruption of the cell output then actuates a motor that opens the door. The basic signal on which the circuit operates is nothing more than a certain d-c current level when the door is closed, and a short duration of zero current when the light beam is broken.

Other applications in which similar types of d-c signals are used include devices for automatically turning on electric lights at night and turning them off at daybreak, and electric eyes used in cameras for automatically adjusting the lens opening.

disadvantages

Steady d-c currents and voltages, as shown on the previous pages, can be made to carry intelligence, and can therefore be used as signals. The signals can be formed by making the dc vary between *zero* and its *full value*, or by having it vary between *two different values*. As far as the signal is concerned, though, both types are the same. The only difference is that the zero reference level is replaced with a d-c reference level.

With the exception of certain applications, such as digital computers, and teletype, d-c signals are not used *extensively* in electronics. There are many reasons for this, with one of the more important being their lack of compatability with many electronic circuits and components. As you learn about the characteristics and uses of a-c signals, the disadvantages of d-c signals will become obvious.

An important point that you should note here is that although d-c *signals* are not used extensively in electronics, d-c *voltages* are *widely used* for supplying *power*. In fact, most of the power supplied to electronic circuits within a piece of equipment is dc. Another important point here is that the d-c signals so far described are signals derived from a *constant* value of dc. *Fluctuating d-c* signals are used extensively in electronics, but their characteristics are so similar to a-c signals that they will be considered as ac.

D-c signals can vary between zero and the steady value of dc

Or they can vary between two different values of dc

Fluctuating d-c signals, which vary in magnitude but not in direction, are widely used in electronics. They are similar in many ways to a-c signals, and will be considered as such

summary

☐ Electronics can be defined as the study of how electricity is used to tell, show, or otherwise inform. To do this, the electricity must carry intelligence, and it is this intelligence that differentiates the study of electricity from the study of electronics. ☐ When an electrical current or voltage carries intelligence, it is referred to as a signal. Intelligence is carried on a signal by means of characteristics such as amplitude, phase, or frequency. ☐ Signals used in a wide variety of circuits and equipments are similar. It is possible, therefore, to divorce signals from specific circuits and equipment and study them from the standpoint of how they carry intelligence and interact with other signals.

☐ Intelligence can be added to a voltage or current to create a signal in many ways. ☐ Probably the simplest way to produce a signal is to interrupt a steady d-c current. The result is a series of pulses when current flows and no pulses when current stops. ☐ If current pulses are made to correspond to some code, they carry intelligence and can be used to transmit messages. The important thing is that the pulses, or interruptions in current, must occur in some logical sequence.

☐ D-c signals are not as widely used as are a-c signals. Nevertheless, they have many important uses, especially in applications involving computers and teletypewriter communications. ☐ Although d-c signals are not used extensively in electronics, d-c voltages are widely used for supplying power to the circuits in electronic equipment. ☐ Fluctuating d-c signals have characteristics very similar to a-c signals.

review questions

1. Name two forms of energy that electricity is used to generate.
2. When is electricity considered to be the field of electronics?
3. Name five fields in which electronics is widely used today.
4. What is a *signal*?
5. Give an example of an electrical signal that you encounter every day.
6. Name three characteristics of a current or voltage that can be varied to produce a signal.
7. Does a 5-ampere current carry more intelligence than a 2-ampere current?
8. Are d-c signals used in electronics?
9. Draw a simple circuit that can produce d-c signals.
10. Can intelligence be added to a fluctuating d-c voltage?

a-c signals

You are now ready to learn about a-c signals. You will see in the rest of this volume how ac, because of its extreme versatility and many desirable characteristics, is used for an almost limitless variety of electronic signals. Before proceeding, though, it is important for you to realize that practically everything you will learn has as its basis the fundamental principles of ac. There is a change in *emphasis,* to be sure, away from the transfer of electrical power and to the transmission or production of intelligence. Nevertheless, the basic concepts of ac are the same, regardless of the specific purpose for which it is being used. Whether it is providing energy to light a city, or carrying the television signal to your home, ac still varies in both magnitude and direction. It is still affected the same way by inductors and capacitors. It still has the basic properties of frequency, phase, amplitude, etc. And it still can be represented graphically by means of waveforms. In other words, a-c electricity is the starting point for learning about a-c electronic signals.

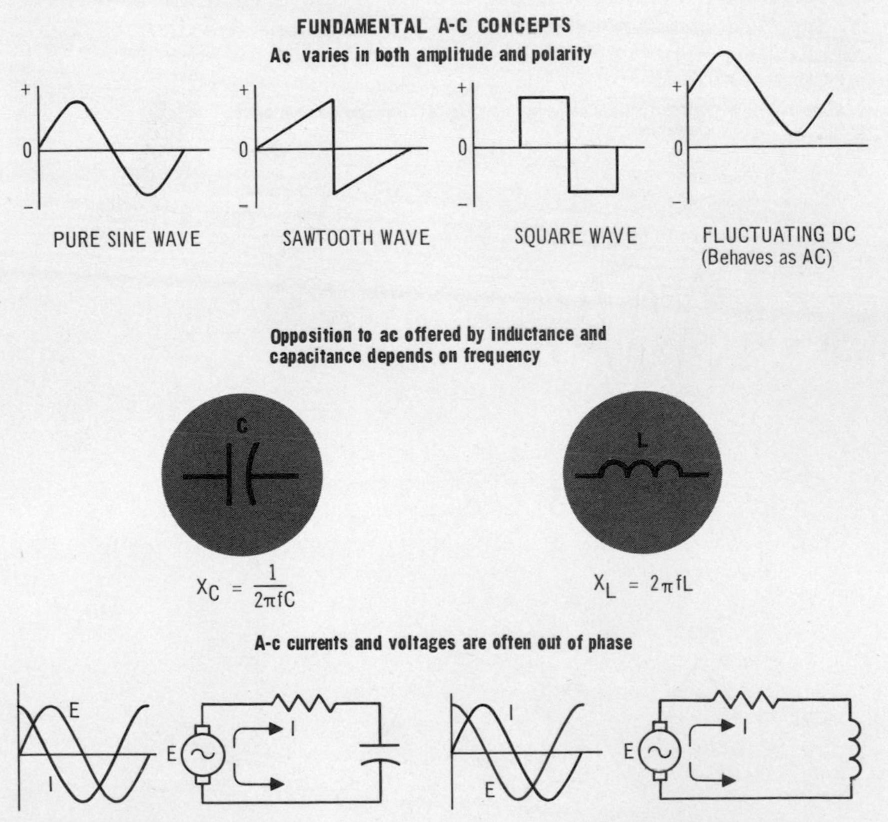

the continuous a-c wave

In electricity, you learned that an a-c wave is made up of a *continuous* succession of sine waves. The sine waves are all *identical* in every way, having the same amplitude, the same period, the same frequency, etc. Because of this absolute similarity, every individual cycle, or sine wave, of a continuous a-c wave looks exactly like every other cycle. Thus, as a carrier of intelligence, the continuous a-c wave is similar to a steady level of dc. It has no *meaningful* variations or changes which, as you know, are necessary for intelligence. Of course, the amplitude of the wave does change. But the changes are exactly the same during each cycle.

You recall that even though a steady value of dc carries no intelligence, it can be broken into a series of pulses that can then represent intelligence. Essentially, the same can be done with a continuous a-c wave. In addition, a-c waves have other characteristics that can be varied, such as the *amplitude* and the *frequency,* as well as some others that are shown. Before describing how intelligence can be inserted on a continuous a-c wave, however, certain basic information will be reviewed on the type and ranges of frequencies commonly used in electronic signals.

Characteristics of continuous a-c waves that are important when they are used as signals are

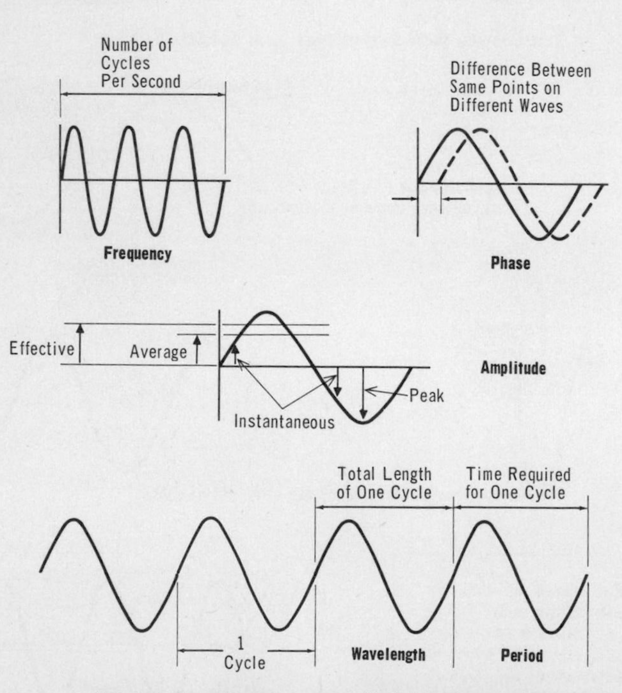

Each cycle of a continuous a-c wave is identical to every other cycle

audio frequencies

The frequency of a continuous a-c wave is the number of times per second that the wave makes a complete cycle from zero, through to its maximum positive value, then through to its maximum negative value, and back to zero. Each full cycle is one complete sine wave. The frequency is normally expressed in cycles per second, so the higher the frequency of a wave, the more times it reverses direction each second. A-c waveforms can be produced having frequencies from as low as a few cycles to billions and even trillions of cycles per second. A cycle per second is called a *hertz*, in honor of the prominent scientist of that name, and is abbreviated *Hz*. The frequency of 10 cycles per second, then, is 10 Hz; 10,000 cycles per second is 10 kHz; and 10,000,000 cycles per second is 10 MHz. MHz stands for megahertz, and kHz stands for kilohertz.

One extremely important band of frequencies corresponds to the frequency range of the *human ear:* those frequencies that people can hear. These frequencies start at about 20 Hz, and extend to about 20,000 Hz. The frequencies in this 20-to-20,000-Hz range are therefore called *audio frequencies*, since they correspond to audible sound.

Often, audio frequencies are also referred to as *sonic frequencies* because they are *sound frequencies*. As a result, frequencies below 20 Hz can be referred to as *subsonic*, and those just above the audio band can be called *supersonic*, *ultrasonic*, or *hypersonic* signals.

This Complex Sound Wave is Made Up of Many Simple Waves

Sound waves are normally complex and are not the simple type of audio sinusoidal wave with which you are familiar

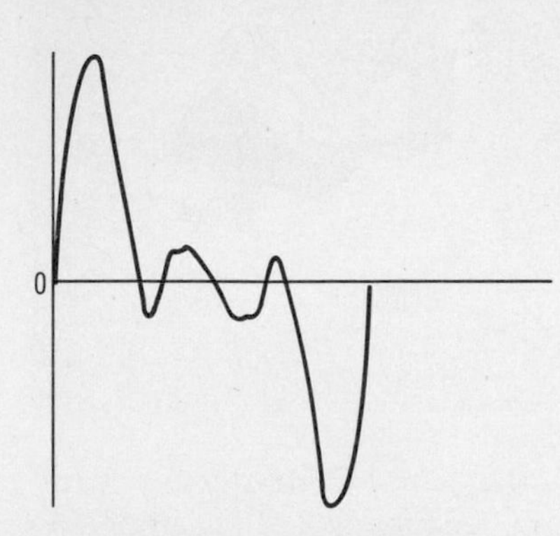

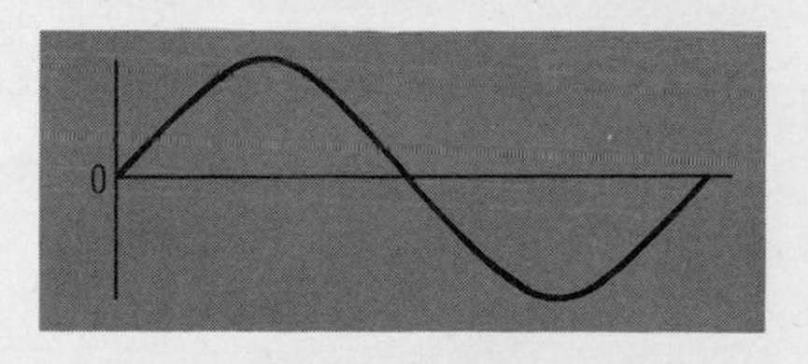

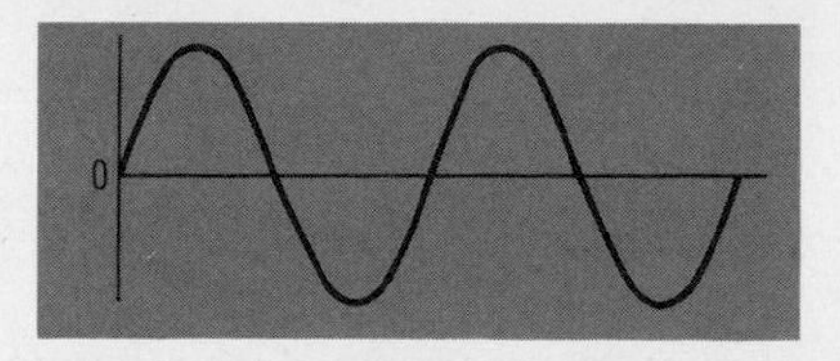

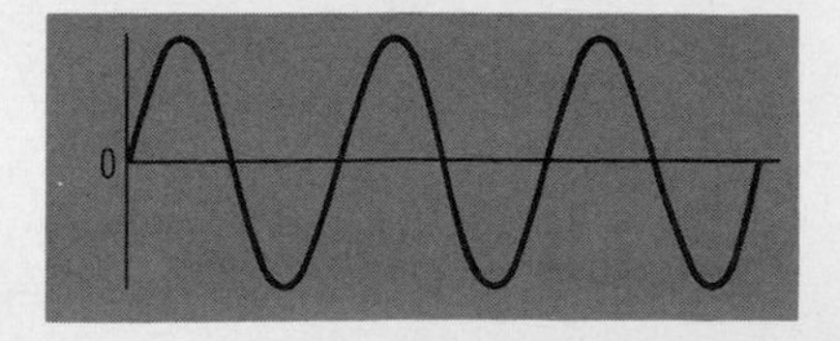

The complex sound waves are made up of many simple sinusoidal audio waves. When the human ear hears a sound represented by a complex sound wave, it is actually sensitive to all the constituent simple sinusoidal waves that are in the audio range

higher frequencies

Signal frequencies as high as thousands of megahertz (MHz) are used in electronics. You already know that those frequencies between 20 and 20,000 Hz are called audio frequencies. They are often, however, also called *very-low frequencies*, since they are the lowest frequencies used for electronic signals. The frequencies above the audio, or very-low, frequencies are also divided into bands, and these bands have designations, such as low-frequency, medium-frequency, etc., which describe the *relative* highness or lowness of the frequencies in the band. Similarly, any frequency in the medium-frequency band is higher than any in the low-frequency band.

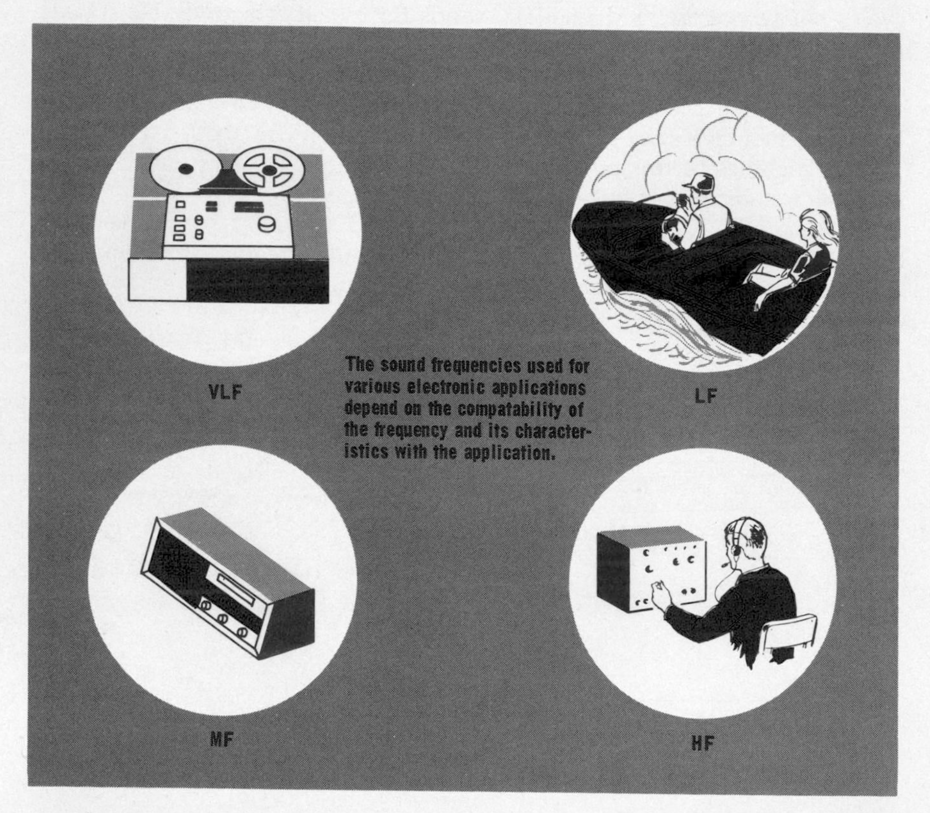

The sound frequencies used for various electronic applications depend on the compatability of the frequency and its characteristics with the application.

The designations of the bands above the very-low frequency band (VLF) are: low frequency (LF), medium frequency (MF), high frequency (HF), very-high frequency (VHF), ultra-high frequency (UHF), super-high frequency (SHF), and extremely-high frequency (EHF). In addition, you have probably heard people refer to *short waves* and *microwaves*. These designations also refer to frequency ranges, but do so in terms of *wavelength*, as meter or centimeter bands. This is possible, since the frequency and wavelength of a wave are directly related.

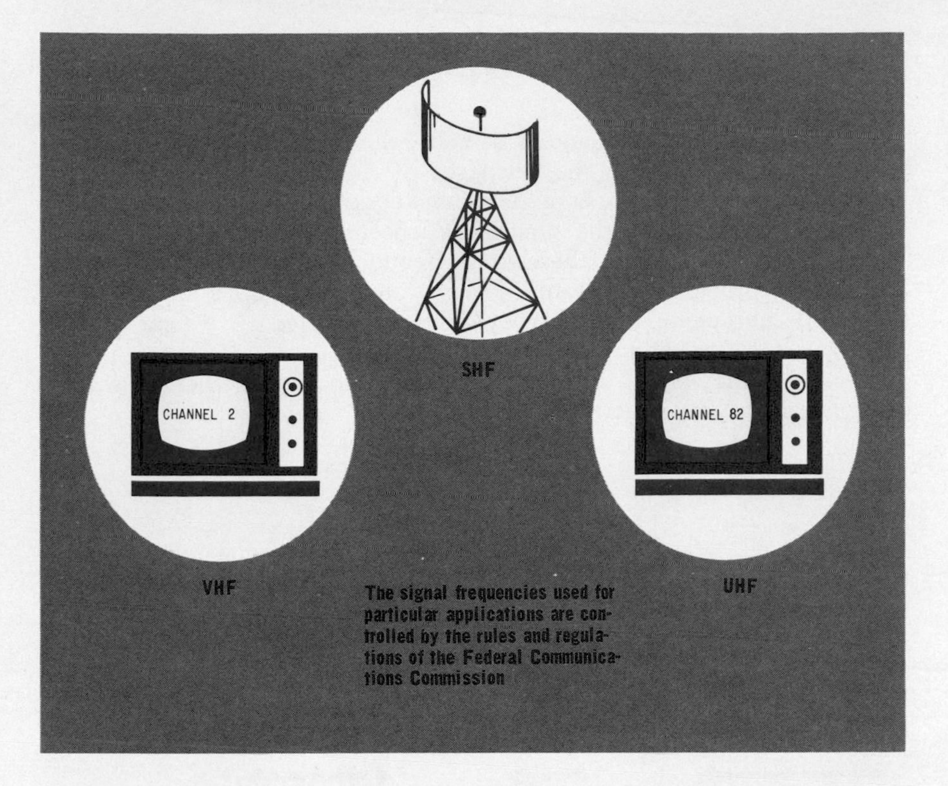

higher frequencies (cont.)

The exact relationship of frequency and wavelength is that the wavelength is equal to the velocity of the wave, which is normally taken as the speed of light, divided by the frequency:

$$\text{Wavelength (meters)} = \frac{\text{speed of light (meters/sec)}}{\text{frequency (Hz)}}$$

$$\lambda = \frac{300{,}000{,}000}{f}$$

With this method, you can say that signals are in the 30-meter band, 50-meter band, and so on; or in the 10-centimeter band, etc. The frequency range covered by the terms short wave and microwave are not exact, with different groups and organizations using different ranges. For this reason, the designations of LF, MF, HF, etc. are preferred, since the more precise definitions are widely accepted. Similarly, in the field of radar, letters such as K, X, and L are often used to designate certain frequency bands. These designations were originally used for purposes of military secrecy, but have now come into general use. Like short waves and microwaves, though, the limits of these bands are not precisely defined.

the frequency spectrum

All of the possible frequencies of continuous a-c waves make up the *frequency spectrum*. For the purposes of electronic signals, the frequency spectrum starts at about 20 Hz and continues up to approximately 30,000 MHz. Actually, the frequency spectrum extends far beyond 30,000 MHz. However, at these high frequencies, a-c waves no longer have the characteristics normally associated with electronic applications and equipment. Instead, they are in the category of heat waves, light waves, X-rays, and so on.

The frequency spectrum of electronic signals, together with the designations of the various frequency bands, is shown.

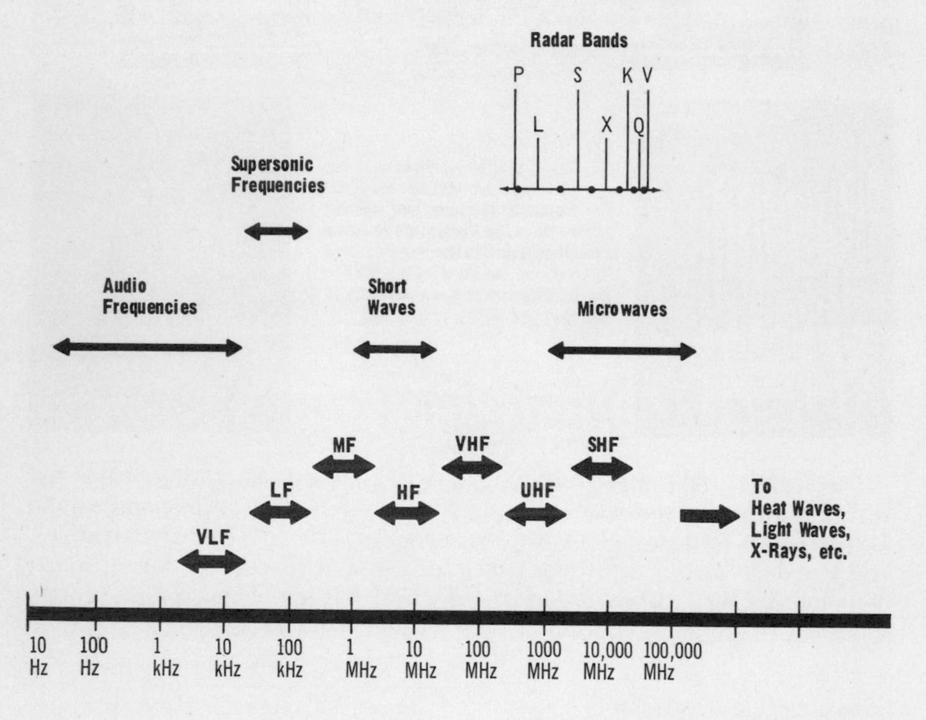

THE FREQUENCY SPECTRUM

Band	Frequency Range
Very-low frequency, VLF	< 3 kHz to 30 kHz
Low frequency, LF	30 kHz to 300 kHz
Medium frequency, MF	300 kHz to 3000 kHz
High frequency, HF	3 MHz to 30 MHz
Very-high frequency, VHF	30 MHz to 300 MHz
Ultra-high frequency, UHF	300 MHz to 3000 MHz
Super-high frequency, SHF	3000 MHz to 30,000 MHz
Extremely high frequency, EHF	Above 30,000 MHz

the interrupted continuous wave

As stated, a continuous a-c wave, like a steady value of dc, carries no intelligence. However, if the wave can be *interrupted* so that it becomes a series of *pulses* that corresponds to some known code, then the wave carries intelligence. Such signals provide *interrupted continuous-wave* (CW) *transmission*, since they do nothing more than go *on* and *off*.

A pulse of a continuous wave actually consists of many cycles of the current or voltage that makes up the wave. However, it is the *presence* of the pulse, or its *duration*, that makes the intelligence, not the number of cycles contained in a pulse. For example, if a one-second pulse represented the letter A, it would make no difference whether a 100-Hz or a 1000-Hz frequency was used.

For the purposes of carrying intelligence, these CW pulses are identical, even though one is derived from an 8-Hz wave and the other from a 2-Hz wave. What makes them the same is that they are both 1-second long

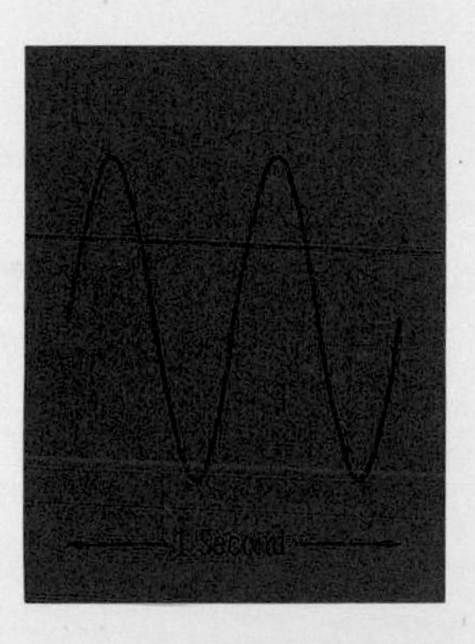

Interrupted CW transmission is widely used in radiotelegraphy, where the continuous wave is broken into pulses that correspond to the familiar dots and dashes of the Morse code. The dots are short pulses, and the dash pulses are three times longer. Each letter of the alphabet then has its own combination of dots and dashes. Thus, pulses corresponding to any combination of dots and dashes can be produced.

In Morse Code, the interrupted CW signal for the letter I (dot – dot) would look like this:

Similarly, the signal for the letter R (dot-dash-dot) would look like this:

summary

□ A continuous a-c wave is made up of a succession of identical sine waves. Such a wave has no meaningful variations, and therefore does not carry intelligence. □ If an a-c wave is broken into a series of pulses that correspond to some known code, it then becomes a signal, since it carries intelligence. A signal of this sort is called an interrupted CW signal. □ Interrupted CW transmission is widely used in radio telegraphy, where the signal interruptions correspond to the dots and dashes of the Morse code.

□ A-c waves can be produced having frequencies from as low as a few Hz to as high as many billions of Hz. □ Frequencies from about 20 to 20,000 Hz correspond to audible sound, and are called audio frequencies. □ All of the possible frequencies of the continuous a-c waves make up the frequency spectrum. For the purposes of electronics, the frequency spectrum is divided into various bands, with each band covering a certain range of frequencies. □ The frequency bands and their designations are: very-low frequency (VLF), low frequency (LF), medium frequency (MF), high frequency (HF), very-high frequency (VHF), ultra-high frequency (UHF), super-high frequency (SHF), and extremely-high frequency (EHF).

□ A-c waves can be described in terms of wavelength as well as frequency. This is because the frequency and wavelength of a wave are directly related. □ Wavelength is equal to the velocity, normally taken as the speed of light, divided by the frequency. As an equation, $\lambda = 300{,}000{,}000/f$; where λ is the wavelength in meters, and f is the frequency in Hz.

review questions

1. What is the simplest way of having a steady continuous wave carry intelligence?
2. What is the wavelength of a 10-MHz a-c wave?
3. How can an interrupted continuous a-c wave be made to carry intelligence?
4. A 50-kHz wave is in what frequency band?
5. Which has a longer wavelength: an audio frequency or a subsonic frequency?
6. Which has the higher frequency: a UHF or a VHF frequency?
7. Which has the shorter wavelength: a UHF or a VHF frequency?
8. Which has the greater amplitude: a UHF or a VHF frequency?
9. What is the frequency of a 1-meter wave?
10. Increasing the amplitude of a continuous a-c wave has what effect on the wavelength?

the modulated continuous wave

You have seen how intelligence can be added to a continuous a-c wave by interrupting it in such a way that it is broken into a series of pulses that correspond to some known code. Although this type of signal is widely used, it has certain disadvantages; the most important being that it cannot easily carry intelligence that corresponds to *sound,* such as the human voice or music. Therefore, when a signal is to be used for carrying sound, other means of adding the intelligence are usually used. In these methods, some characteristic of the *continuous* a-c wave is *controlled* by the sound in such a way that it varies in exactly the same manner as the sound wave. The continuous a-c wave is then said to be *modulated* by the sound.

Modulation is a process whereby some characteristic of a wave is varied by another wave

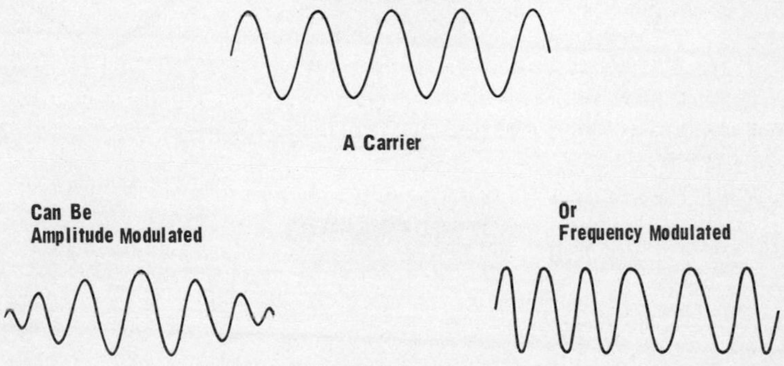

A Carrier

Can Be Amplitude Modulated

Or Frequency Modulated

Two of the most commonly used types of modulation are amplitude and frequency modulation

Both are widely used for inserting sound, or audible, intelligence on carrier waves

The two most common characteristics of an a-c wave that are modulated by sound are the *amplitude* and the *frequency.* When the sound controls the amplitude of the wave, it is *amplitude modulation;* and when the frequency is controlled by the sound, it is *frequency modulation.* The a-c wave is called the *carrier,* since it is made to "carry" the sound intelligence.

In actual practice, amplitude and frequency modulation are used to add many other types of intelligence, besides sound, to a carrier wave. However, the principles are the same regardless of the type of intelligence involved, as you will learn.

amplitude modulation

When the *amplitude* of one wave is varied in accordance with another wave that represents some form of intelligence, the process is called amplitude modulation. The wave being modulated is the *carrier*, and the other is the *modulating signal* or wave. In amplitude modulation, the *peak-to-peak* amplitude of the a-c carrier is varied with the intelligence; the carrier then consists of sine waves whose amplitudes follow the amplitude variations of the modulating wave, so that the carrier is contained in an *envelope* formed by the modulating wave.

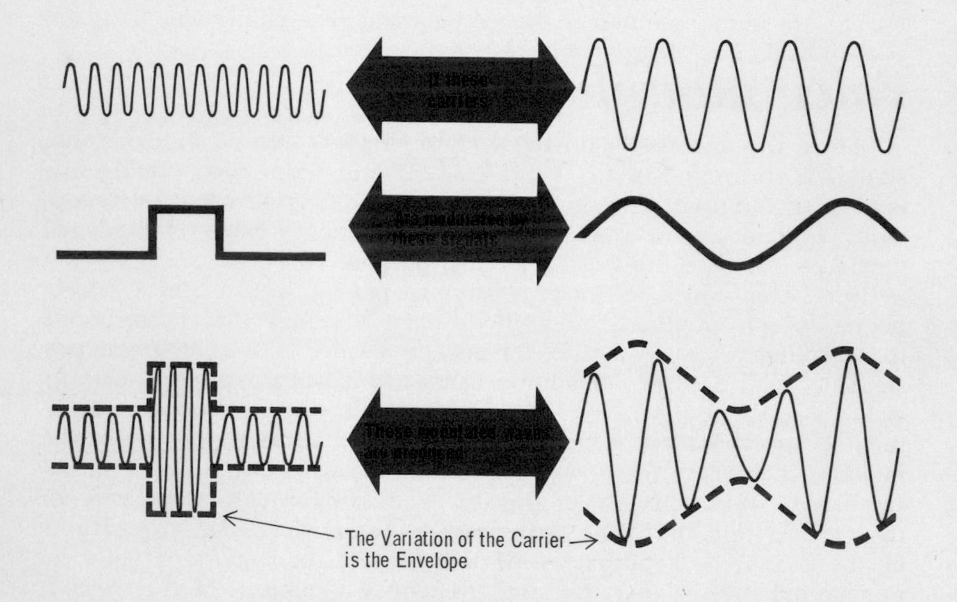

If the carrier and the lower half of the envelope could be removed from a modulated wave, the remaining upper half of the envelope would exactly duplicate the modulating wave, which represents the intelligence being carried. This is done to recover, or remove, the intelligence from the modulated wave. The process is called DEMODULATION, and will be covered later

You may now wonder why a signal that already carries intelligence (the modulating wave) should be used to modulate another wave (the carrier). As you will learn, radio transmission of higher frequency signals is *cheaper* and more *reliable* than transmission of very low-frequency signals. You need a lot of power to transmit a low-frequency signal, and it cannot be transmitted very far. Thus, low-frequency intelligence is placed on, or modulates, a higher frequency signal, which is often called a *radio-frequency* (r-f) *carrier* since it is suitable for radio transmission. The term "radio frequency" does not define any specific frequency band. It just means that the frequency is high enough for practical radio transmission.

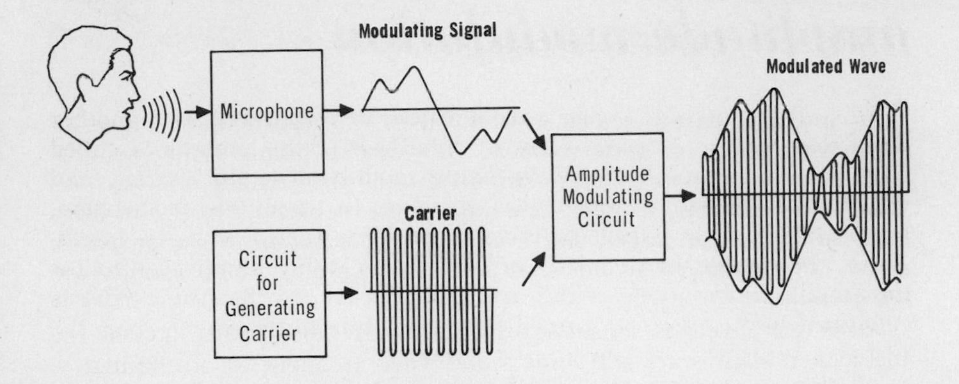

voice modulation

One of the most common types of intelligence carried by electronic signals is the sound of the *human voice*. And when such intelligence is to be transmitted by radio, an electrical voice signal is first produced, which is then used to modulate a higher frequency carrier. Amplitude modulation is used most often for this purpose.

In a typical voice modulation setup, a person speaks into a microphone, which produces an audio-frequency signal that corresponds to the sound waves created by the speaker's voice. The audio-frequency signal is then used to modulate a carrier that has a frequency usually well above the audio range. After modulation, the peak-to-peak amplitude of the carrier varies according to the modulating signal. Thus, the modulated wave varies in two ways: one of these is the *rapid* (high-frequency) variations of the instantaneous amplitude and polarity of the carrier; and the other is the *slower* (audio-frequency) variations of the peak-to-peak amplitude of the carrier. An important point to understand here is that the audio-frequency variations of the carrier, which actually represent the sound intelligence being transmitted, vary in both frequency and amplitude. The frequency variations represent the pitch of the sound, and the amplitude variations, the loudness of the sound. The carrier on the other hand, only varies in amplitude. Its frequency does not change.

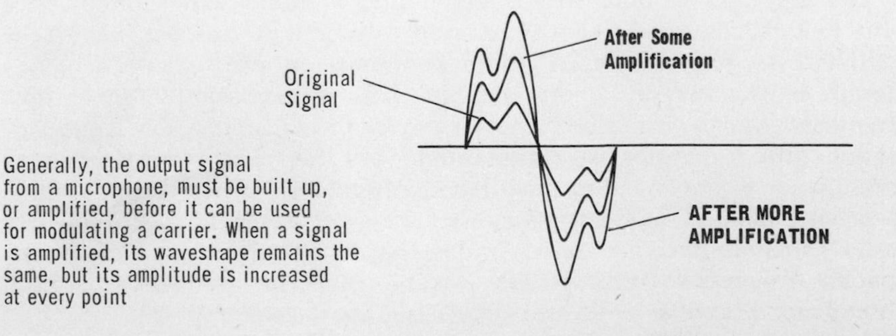

Generally, the output signal from a microphone, must be built up, or amplified, before it can be used for modulating a carrier. When a signal is amplified, its waveshape remains the same, but its amplitude is increased at every point

usable voice range

As you learned previously, sound waves, and particularly those of the human voice, are very *complex*. They contain many simple sounds. It is the combination of these simple sounds that makes one person's voice different from another, and allows the voice to express feelings such as sorrow or anger. For basic *voice communications* purposes, many of the simple sounds contained in complex voice sounds are *unnecessary*.

Although the ear is sensitive to sounds from about 20 to 20,000 Hz, the spoken sounds are still understood when frequencies are limited to about 200 to 2700 Hz. Thus, electronic signals do not always have to duplicate the sound waves. This often makes it possible to reduce the complexity and cost of the electronic circuits used for sound reproduction and transmission. Of course, in certain areas, such as the high-fidelity reproduction of music, the electronic sound signal must duplicate closely the original sound waves. For standard AM radio broadcasting, on the other hand, audio frequencies above 5000 Hz are eliminated, yet the quality of the sound is satisfactory for most purposes.

First Person

Second Person

If two people pronounced the letter O, the waveforms of their sounds would be similar and yet different in some respects

The differences arise from the fact that, although in both cases the sound O was recognizable, the two waveforms are made up of a somewhat different combination of frequencies

This Sound Wave MIGHT BE REPRODUCED INTELLIGIBLY BY **This Electronic Sound Signal**

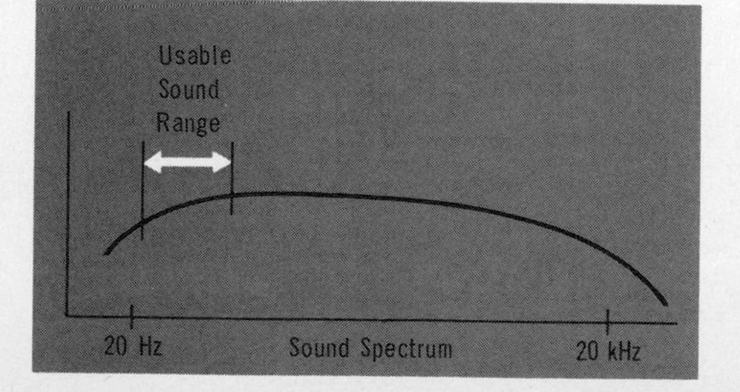

Some of the simple component frequencies of a complex sound wave can be omitted and the sound will still be intelligible. When frequencies are removed, the waveform of the sound changes

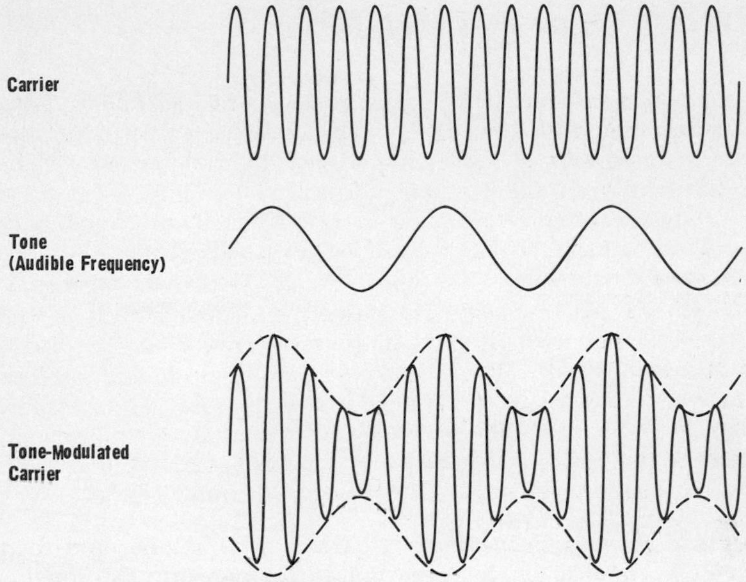

If a continuous-wave carrier is amplitude modulated by an audible frequency, the result is a tone-modulated carrier. Such a signal can carry intelligence in various ways

tone modulation

Previously, you learned about interrupted CW signals. These consist of pulses that conform to the dots and dashes of the Morse code, and which we made by interrupting a continuous a-c wave. Although interrupted CW signals provide a good means for transmitting intelligence, they are subject to certain conditions that can render them unintelligible. The reason for this is that the on–off conditions of the carrier are really only d-c levels, which must be changed to a signal you can hear at the receiver. Sometimes, because of *instabilities* in the electronic circuits and equipment used, the CW signal will change, or *drift*, in frequency. The circuits that convert the signal to audible sounds then can no longer perform their function, and the signal is, in effect, lost.

One way of overcoming this problem is by amplitude modulating the continuous wave with an *audible* signal. Sinusoidal signals of 500 or 1000 Hz are frequently used for this. When the continuous wave is modulated in this way, it can still be interrupted to form dot-and-dash pulses. However, now each pulse contains an audio-frequency envelope, and the dot-and-dash sounds are already in the signal. So, to convert the pulses to audible sounds, all that is required is for the carrier and one-half of the envelope to be removed in some way. Circuits for accomplishing this are relatively simple and reliable.

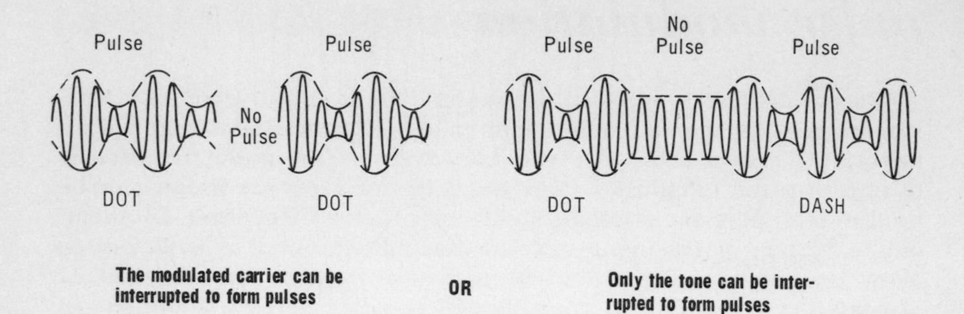

The modulated carrier can be interrupted to form pulses

Only the tone can be interrupted to form pulses

tone modulation (cont.)

It is also possible to modulate a carrier with a tone, and then just *interrupt the tone* to produce the pulses for carrying the intelligence. The *no-pulse* condition is then represented by the *unmodulated carrier*. With this method, the signal is continuous and only the intelligence portion is interrupted. In radio transmission, as you will learn later, this method makes it possible to *maintain radio contact* even when no intelligence is being sent.

Tone modulation can also be used for the methods in which more than a single tone modulates a carrier. In these methods, one tone can correspond to a pulse and another to no pulse; or one tone can be a dot and another tone can be a dash; or the various tones can even represent different intelligence. Of course, when multiple tones are used, they must have different frequencies so that they can be distinguished from each other.

Sometimes a single *continuous tone* is used to modulate a *continuous-wave carrier*. In this case, the modulated wave itself carries *no* intelligence, but the way in which the wave is used does. You will learn more about the uses of this type of signal later.

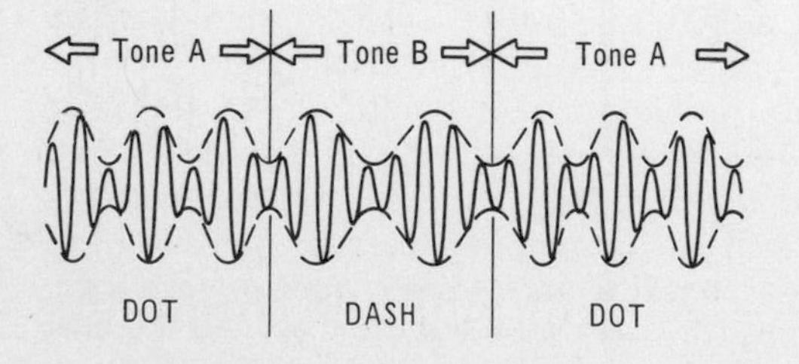

More than one tone can be used

A single tone can continuously modulate the carrier

pulse modulation

When intelligence is to be transmitted by radio at UHF and SHF frequencies, the power requirements of the electronic equipment is an important consideration. One way of decreasing the power required is to break up the intelligence into *small bits* or *samples,* which can be used to reproduce the original intelligence after transmission. In effect, only a *portion* of the intelligence is transmitted, but it is sufficient to allow the *total* intelligence to be recreated from it. One method of accomplishing this type of transmission is called *pulse modulation.* In this type of modulation, the modulating signal is first converted to a series of pulses whose amplitudes correspond to the instantaneous amplitudes of the modulating signal. These *amplitude-modulated pulses* are then used to amplitude modulate a carrier wave, with the result that the modulated carrier consists of a series of pulses having amplitudes that correspond to the intelligence being transmitted.

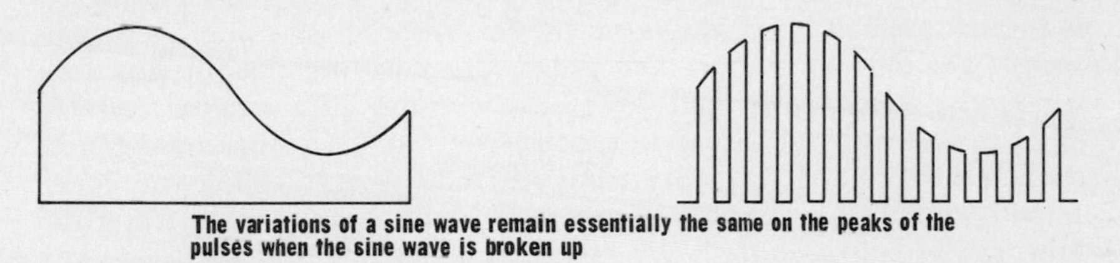

The variations of a sine wave remain essentially the same on the peaks of the pulses when the sine wave is broken up

Actually, the type of pulse modulation just described is called *pulse amplitude modulation* (PAM). There are other methods of pulse modulation frequently used. These other methods, as well as a more detailed description of PAM, are covered later.

For an identical carrier and modulating signal the peak and minimum transmitted power is the same whether continuous or pulse modulation is used. However, since with the pulse method there are periods when the signal is not being transmitted at all, the average power of the pulse signal is much less than that of the continuous one.

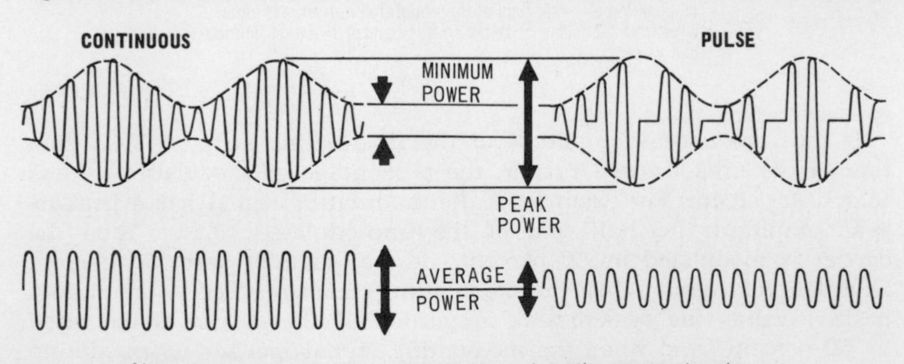

Although a continuous and a pulse signal may have the same peak and minimum power, the average power of the pulse signal is much lower

percentage of modulation

In the discussion of amplitude modulation, nothing has been said about the *relative amplitudes* of the modulating signal and the unmodulated carrier. The relationship between these two amplitudes is the *percentage of modulation,* which expresses the extent to which the carrier is modulated. When the peak-to-peak amplitude of the modulating signal *equals* the peak-to-peak amplitude of the unmodulated carrier, the carrier is said to be *100 percent* modulated. Thus, in the case of 100 percent modulation, the modulating signal goes far enough positive to double the peak-to-peak amplitude of the carrier, and far enough negative to reduce the peak-to-peak amplitude of the carrier to zero.

100 percent modulation occurs when the peak-to-peak amplitudes of the modulating signal and the unmodulated carrier are equal

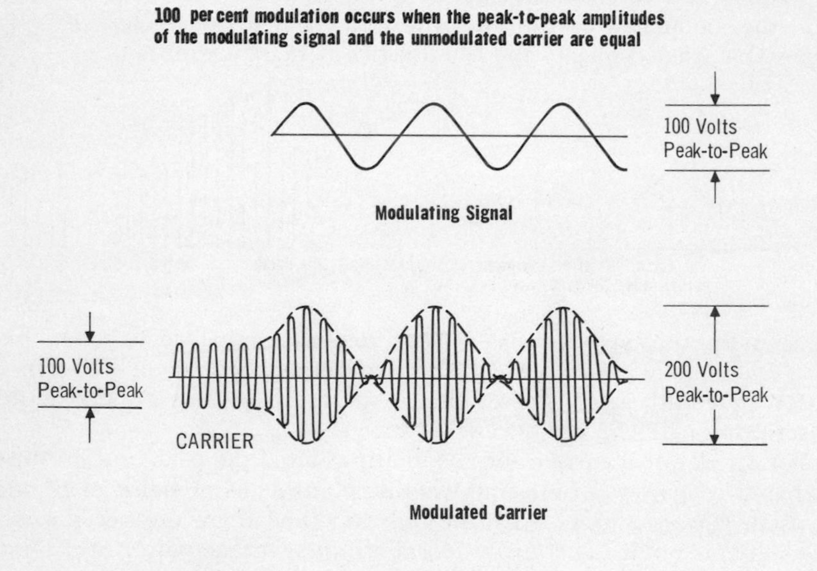

The peak-to-peak amplitude of the modulated carrier, therefore, varies from a minimum of zero to a maximum of twice its unmodulated value

If the peak-to-peak amplitude of the modulating signal is *less* than that of the unmodulated carrier, the percentage of modulation is less than 100 percent. For example, if the modulating signal has a peak-to-peak amplitude one-half that of the unmodulated carrier, then the carrier is modulated by 50 percent; or, put another way, 50 percent modulation is used. When the modulating signal reaches its maximum positive value, the peak-to-peak amplitude of the carrier is increased by 50 percent. And when the modulating signal reaches its maximum negative value, the carrier's peak-to-peak amplitude is decreased by 50 percent.

67 percent modulation occurs when the peak-to-peak amplitude of the modulating signal is two-thirds, or 67 percent, that of the unmodulated carrier

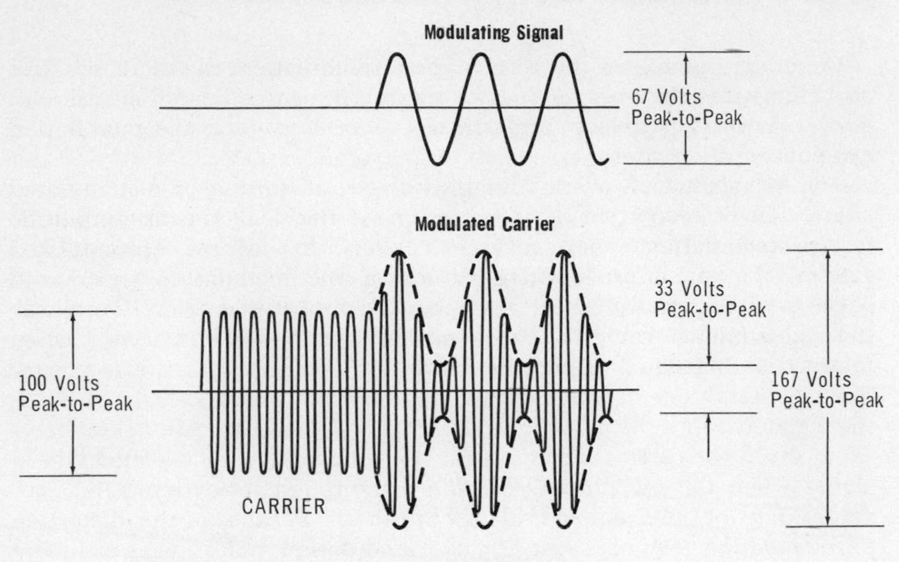

The peak-to-peak amplitude of the modulated carrier, therefore, varies from a minimum value 67 percent less than its unmodulated value, to a maximum 67 percent greater than its unmodulated value

percentage of modulation (cont.)

There are various formulas for calculating the percentage of modulation of an amplitude-modulated carrier. One of the easiest is

$$\text{Percentage of modulation} = \frac{E_{MAX} - E_{MIN}}{E_{MAX} + E_{MIN}} \times 100$$

where E_{MAX} is the greatest peak-to-peak amplitude of the modulated carrier, and E_{MIN} is the smallest. For example, if the peak-to-peak values of a modulated carrier vary from a maximum of 167 volts to a minimum of 33 volts, the percentage of modulation is

$$\begin{aligned}\text{Percentage of modulation} &= \frac{167 - 33}{167 + 33} \times 100 \\ &= 0.67 \times 100 \\ &= 67\%\end{aligned}$$

The equation can only be used for modulation of 100 percent or less. Modulation greater than 100 percent is called *overmodulation,* and is generally undesirable.

overmodulation

Generally, the higher the percentage of modulation, the more effective and efficient is the transmission of an amplitude-modulated signal. But *overmodulation* is usually undesirable, since it *distorts* the modulation envelope of the carrier.

The way in which overmodulation causes distortion of a modulated signal can be seen from the illustration. If the peak-to-peak amplitude of the modulating signal is greater than that of the unmodulated carrier, the maximum positive values of the modulating signal will cause peak-to-peak values of the modulated carrier greater than twice the unmodulated value. There is nothing wrong with this. But when the modulating signal reaches its maximum negative values, it is greater than the amplitude of the unmodulated carrier. This cancels, or *cuts off,* the carrier, since its peak-to-peak amplitude cannot fall below zero. As a result, the carrier variations do not follow those of the modulating signal when the modulating signal goes through its extreme negative values. Part of the negative cycle is *clipped* off. Because of this distortion of the modulated signal caused by overmodulation, percentages of modulation greater than 100 percent are seldom used. Such overmodulation is not used in voice communication systems because the voice will sound clipped or garbled.

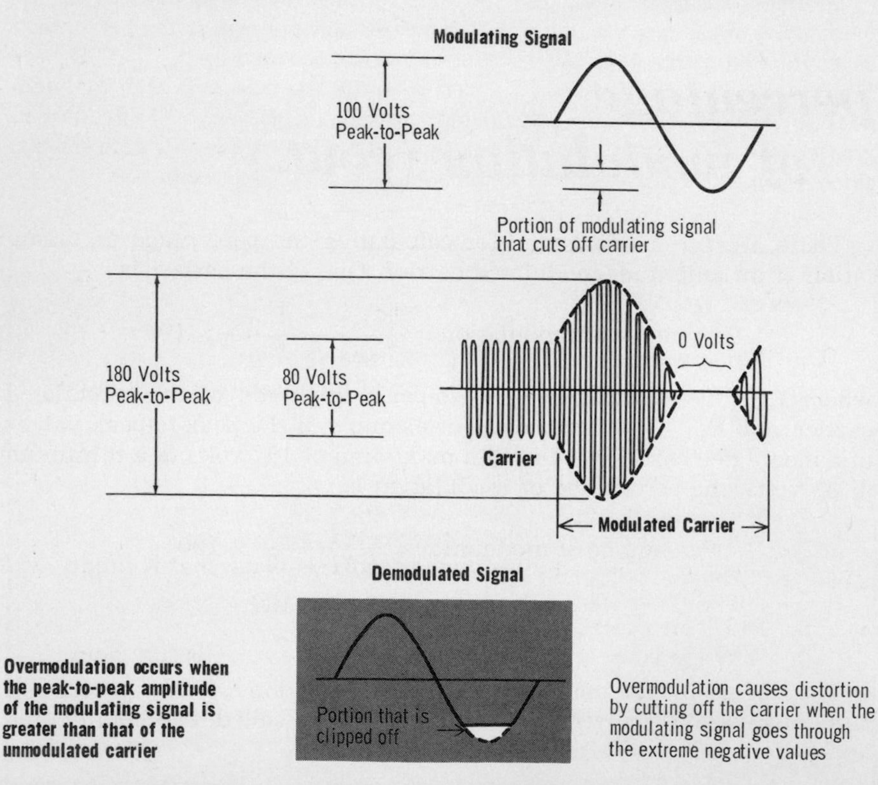

Overmodulation occurs when the peak-to-peak amplitude of the modulating signal is greater than that of the unmodulated carrier

Overmodulation causes distortion by cutting off the carrier when the modulating signal goes through the extreme negative values

summary

□ When sound or other intelligence controls some characteristic of an a-c wave, the wave is said to be modulated. If the amplitude of the wave is controlled, it is amplitude modulated; and if the frequency is controlled, it is frequency modulated. □ An a-c wave being modulated is called the carrier. The signal that is adding the intelligence is called the modulating signal.

□ In amplitude modulation the peak-to-peak amplitude of the carrier is varied in accordance with the intelligence. After being modulated, the carrier is contained in an envelope formed by the modulating wave. □ Simple audio tones, as well as complex voice signals, can be amplitude modulated onto a carrier wave. Tones can carry intelligence by being interrupted to form the dots and dashes of Morse code. □ Before intelligence is modulated onto a carrier, the intelligence can be broken into small bits or samples that can be used to reproduce the original intelligence after transmission. This is called pulse modulation. □ In one type of pulse modulation, called pulse amplitude modulation (PAM), the modulating signal is converted to a series of pulses whose amplitudes correspond to the instantaneous amplitudes of the modulating signal. □ The advantage of pulse modulation is that it requires less average power to transmit a given amount of intelligence.

□ In amplitude modulation, the relationship between the amplitudes of the modulating signal and the unmodulated carrier is expressed as the percentage of modulation. Percentage of modulation can be calculated by $(E_{MAX} - E_{MIN})/(E_{MAX} + E_{MIN}) \times 100$; where E_{MAX} and E_{MIN} are the maximum and minimum peak-to-peak values of the modulated carrier, respectively. □ Modulation greater than 100 percent is called overmodulation and is generally undesirable, since it can cause distortion of the intelligence carried by the signal.

review questions

1. When is an a-c wave considered a carrier?
2. Draw and label the components of an amplitude-modulated signal.
3. What is *tone modulation*?
4. What is *pulse amplitude modulation*?
5. Why is pulse amplitude modulation used?
6. What is *overmodulation*, and why is it undesirable?
7. Draw and label an amplitude-modulated wave that is modulated by 50 percent.
8. Draw an overmodulated wave.
9. Why are voice signals modulated onto a carrier before being transmitted?
10. How does a tone-modulated signal carry intelligence?

side bands

Until now we have been interested in the modulated carrier that results when the amplitude of a carrier is varied by a modulating signal. You have learned that the modulated carrier is a constant-frequency waveform whose amplitude changes in accordance with the modulating signal. Mathematically, though, it can be proven that the modulated carrier consists of *other frequencies* in addition to the carrier frequency. These frequencies are generated as a result of the modulation process, and make up what are known as *side-band frequencies*. Side-band frequencies are not merely mathematical abstractions. They can actually be separated from the modulated carrier, and form the basis for a widely used system of radio transmission.

When this signal . . .

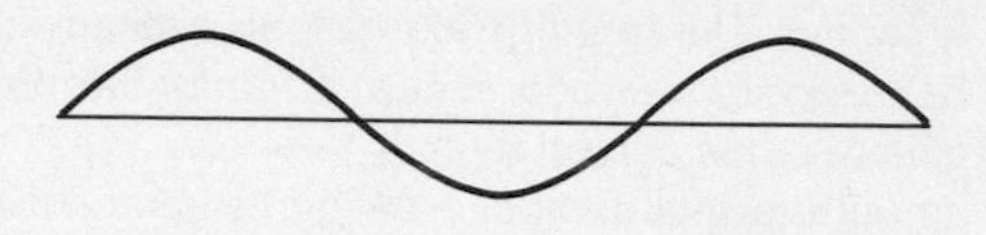

. . . Modulates this carrier . . .

. . . This modulated carrier is produced, and consists of the following components:

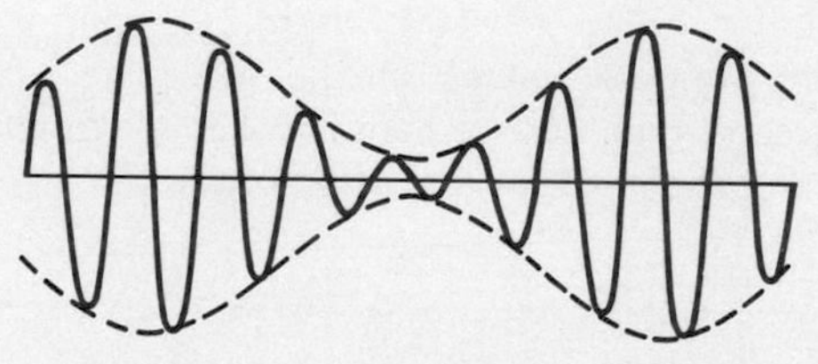

CONSTANT-AMPLITUDE,
UPPER SIDE-BAND FREQUENCY

CONSTANT-AMPLITUDE
CARRIER FREQUENCY

CONSTANT-AMPLITUDE,
LOWER SIDE-BAND FREQUENCY

The amplitude of the modulated carrier at every instant is equal to the algebraic sum of the instantaneous amplitudes of its three components

side bands (cont.)

Side-band frequencies can best be understood by first considering a carrier modulated by a simple sinusoidal modulating signal. The modulating process produces two entirely new frequencies. This results from a process called *heterodyning*, which you will learn about later. One of the new frequencies is equal to the *sum* of the carrier and modulating signal frequencies, and is called the *upper side-band frequency* since it is above the carrier in frequency. The other new frequency equals the *difference* between the carrier and the modulating signal frequencies, and is called the *lower side-band frequency*.

The modulated carrier consists essentially of the two side-band frequencies and the carrier frequency, all having *constant amplitudes*. When the waveforms of these three component frequencies are added, they produce the waveform of the modulated carrier. Thus, the effect of the side-band frequencies is to provide the *amplitude variations* for the carrier frequency.

As an example of the generation of the side-band frequencies, consider a 30-kHz carrier modulated by a 1 kHz signal. The upper side-band frequency would be 30 + 1, or 31 kHz, and the lower side-band frequency, 30 − 1, or 29 kHz. The amplitude-modulated carrier, therefore, consists of a 29-kHz, a 30-kHz, and a 31-kHz component, each having a constant amplitude.

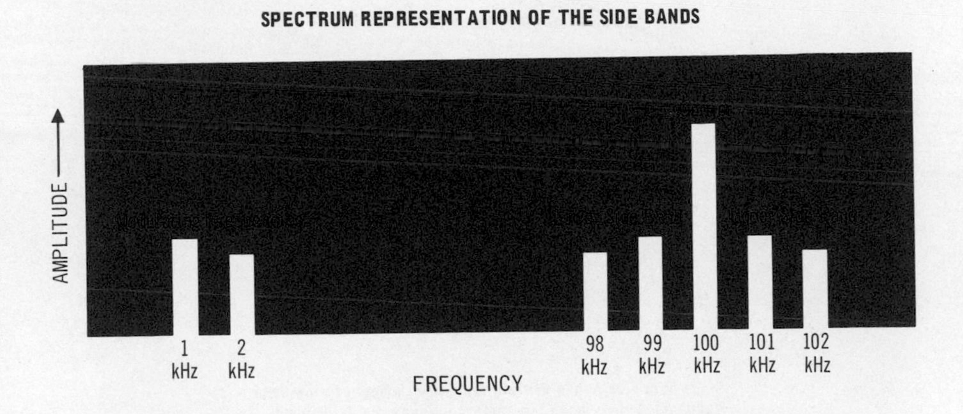

You have seen that when a carrier is amplitude modulated by a single frequency, two side-band frequencies are produced. When a modulating signal consists of more than one frequency, such as in voice modulation, two side-band frequencies are produced for *every* frequency contained in the modulating signal. Thus, if the modulating signal contains two frequencies, four side-band frequencies are produced: two higher than the carrier frequency, and two lower than it. Similarly, if there are ten frequencies in the modulating signal, twenty side-band frequencies are produced; ten on each side of the carrier frequency.

side bands and bandwidth

You can see, then, that there are always *exactly* the same number of side-band frequencies *higher* than the carrier frequency as there are *lower* than it. Together, all of the side-band frequencies above the carrier frequency make up the *upper side band.* In a similar manner, all those below the carrier frequency make up the *lower side band.* The frequencies in the *upper* side band represent the *sum* of the individual modulating frequencies and the carrier frequency, while those in the *lower* side band represent the *difference* between the modulating frequencies and the carrier frequency.

The side-band nature of an amplitude-modulated carrier can be depicted in two ways. One is by use of a spectrum diagram, which shows the amplitude and frequencies of the modulation components in a graphical way (see page 1-29). The other is a pictorial representation of the component waveforms.

PICTORIAL REPRESENTATION OF SIDE BANDS

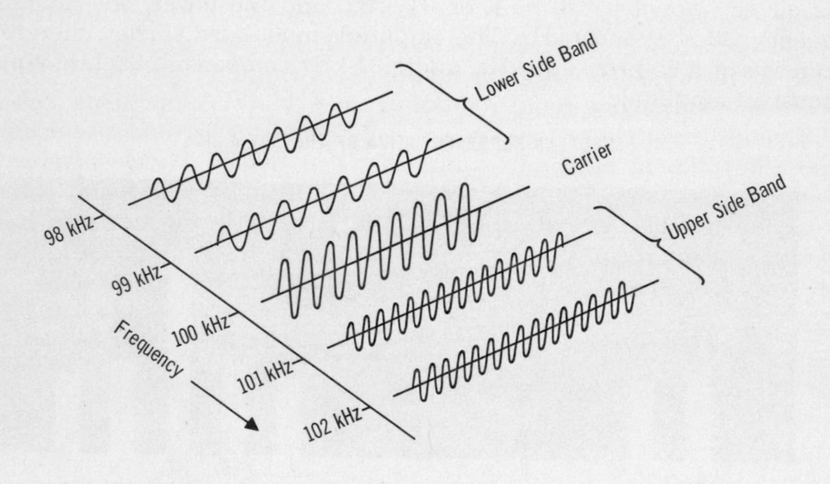

Since both side bands contain the same number of side-band frequencies, they have the same frequency width, or bandwidth

The *bandwidth* of an amplitude-modulated signal is the *frequency range* between the lowest, lower side-band frequency and the highest, upper side-band frequency. Since both of the extreme side-band frequencies are produced by the same modulating frequency, namely the *highest,* the bandwidth can be expressed in terms of the highest modulating frequency. The exact relationship is that the bandwidth is equal to *twice* the highest modulating frequency. For example, if the highest modulating frequency is 3 kHz, the bandwidth is 6 kHz.

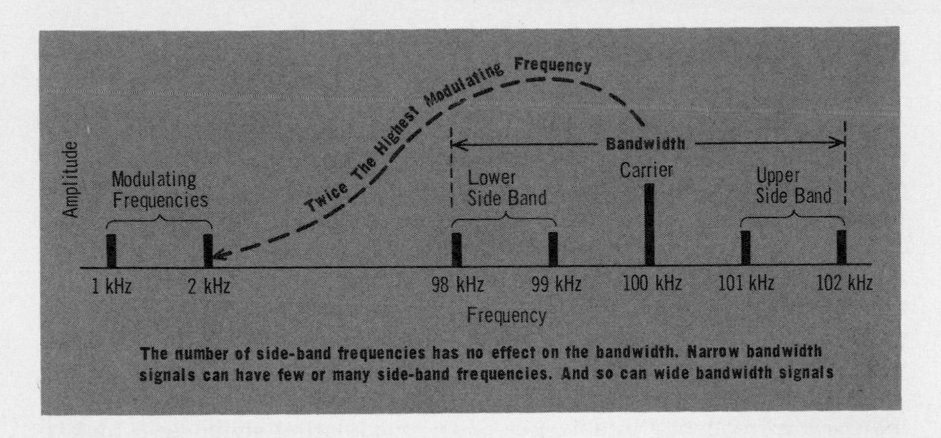

The number of side-band frequencies has no effect on the bandwidth. Narrow bandwidth signals can have few or many side-band frequencies. And so can wide bandwidth signals

bandwidth

The bandwidth of a signal is important for a variety of reasons. One of these is that it is determined by the frequencies in the intelligence being carried by the signal; so if the intelligence is to be carried with no distortion, all of the electronic circuits that process the signal must be able to do their job over the entire bandwidth. From knowledge of inductors and capacitors, you know that these components behave differently at different frequencies. This should give you some idea of the difficulties in building circuits that can handle a wide bandwidth with no distortion. For example, circuits used for processing signals carrying intelligence consisting of only a 1000-Hz tone (2-kHz bandwidth) can be much simpler than circuits that have to process a range of voice frequencies from say 50 to 15,000 Hz (30-kHz bandwidth).

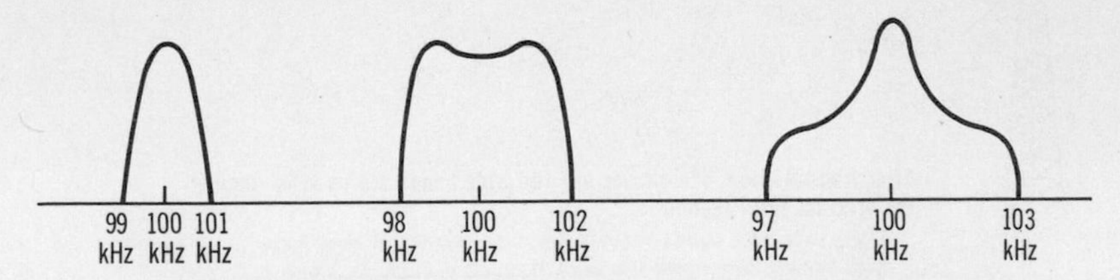

Bandwidth or bandpass curves such as these are usually used to show the side-band frequencies that go with the carrier. The shape of the curve depends on the amplitudes of the side bands

To simplify the circuit requirements, therefore, it is common practice to keep the bandwidth of signals as *narrow* as possible without destroying the intelligibility of the information being carried by the signal. This is why in voice modulation many of the higher audio frequencies are often *eliminated* before modulation and so not included in the signal. Although this introduced some distortion in the signal, it does not destroy the basic information.

side bands and intelligence

You know that when all of the *instantaneous amplitudes* of the upper side-band frequencies, the lower side-band frequencies, and the unmodulated carrier are *added*, the result is the modulated carrier. This shows that it is the side bands that cause the amplitude variations of the carrier, which you know represent the intelligence. In other words, then, the *side bands carry the intelligence*. This may be hard to visualize at first, but it is a fact, and it is useful, therefore, to think of the modulated carrier wave not as a single wave, but as an unmodulated carrier wave with upper and lower side-band waves. When you think in these terms, the purpose of the unmodulated carrier is to serve as a means of *increasing* the low-frequency intelligence of the modulating signal to a higher frequency; namely the side-band frequencies. In the demodulation process, the carrier also provides the means for converting the intelligence contained in the side bands back to its original lower (usually audio) frequency form. Thus, the only purpose of the carrier is to convert the intelligence from one frequency to another.

By means of a carrier, audio-frequency intelligence is converted to high frequencies in the form of upper and lower side bands

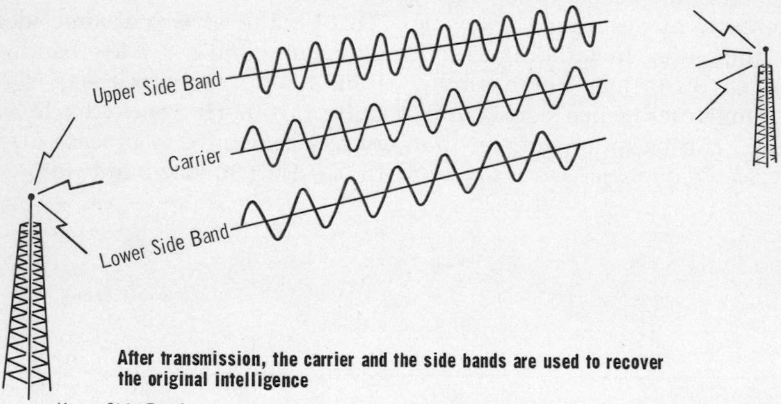

Upper Side Band
Carrier
Lower Side Band

After transmission, the carrier and the side bands are used to recover the original intelligence

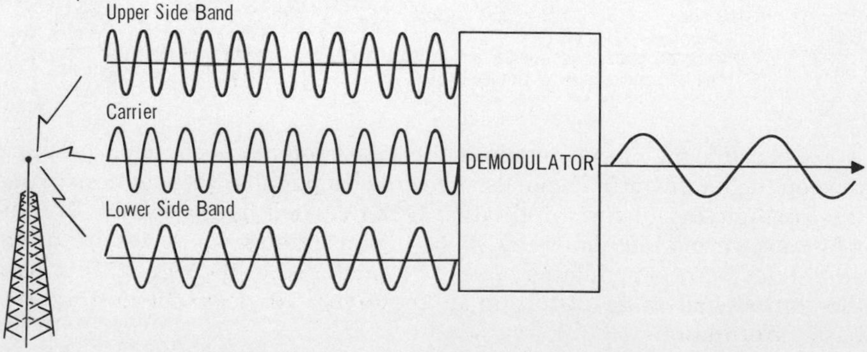

Upper Side Band
Carrier
Lower Side Band
DEMODULATOR

the intelligence of a single side band

If you carefully inspect the waveforms of the side bands and unmodulated carrier, you will see that the addition of only one of the side bands and the unmodulated carrier will produce a waveform that has the same relative amplitude variations that would result if both side bands and the carrier were added. With only one side band, the peak-to-peak amplitudes of the resulting wave, and therefore the modulation envelope, are smaller, but the amplitude *variations* follow the intelligence of the modulating signal. This means, therefore, that the two side bands each contain the entire signal intelligence. As a result, all that is really necessary to recover the intelligence is the unmodulated carrier and one of the side bands, either the upper or lower.

An interesting point here concerns the intelligence carried by radio signals in the commercial AM broadcast band. To conserve bandwidth, these signals are limited by the FCC to a bandwidth of 10 kHz. But since both side bands carry the same intelligence, the actual signal intelligence is limited to a maximum frequency of 5 kHz. The other 5 kHz of the bandwidth has the same intelligence, only at different frequencies.

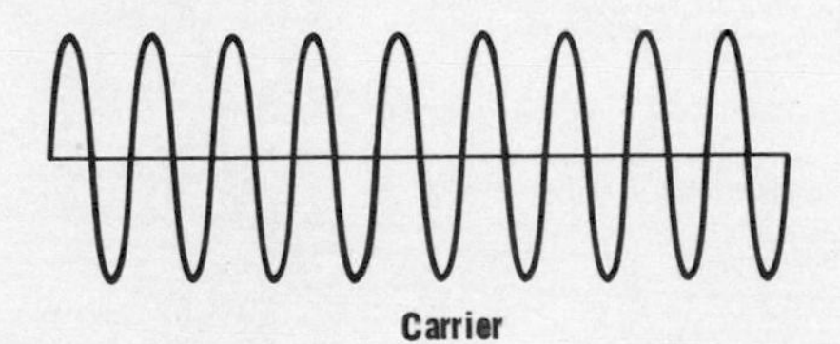

Carrier

Lower Side Band

Upper Side Band

Intelligence Envelope

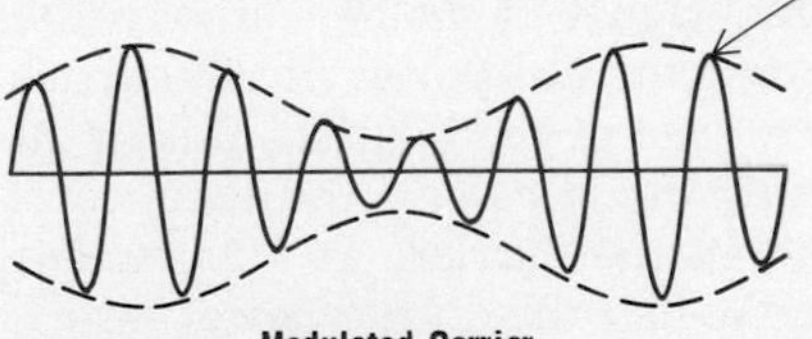

Modulated Carrier

If only the lower side band combines with the unmodulated carrier, the amplitude variations of the carrier are similar, although smaller, than when both side bands are used

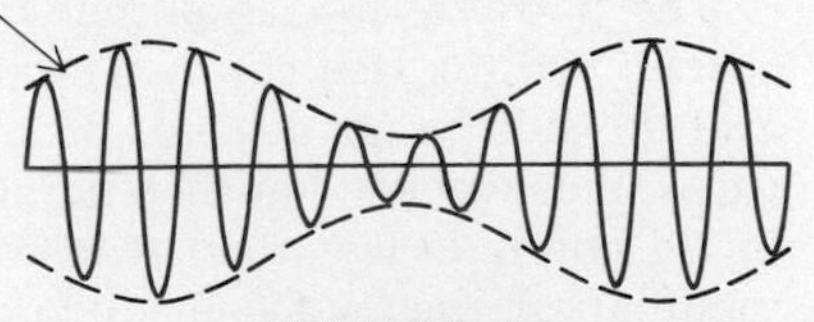

Modulated Carrier

If only the upper side band combines with the unmodulated carrier, the amplitude variations of the carrier are also similar to those produced when both side bands are used

side-band power

Every electronic signal has a certain amount of electrical *energy*. Depending on the nature of the signal, its electrical energy may be measured in terms of voltage, current, or power. The energy of amplitude-modulated signals is usually expressed as *power*, and it determines to a large extent how far the signal can be sent, as well as how much buildup, or amplification, will be required to increase the signal to a usable level after transmission.

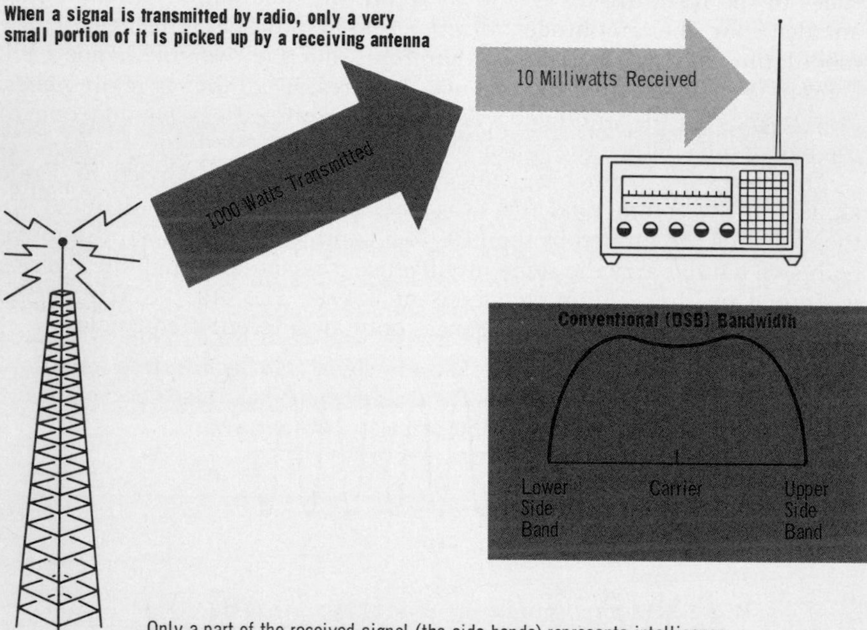

When a signal is transmitted by radio, only a very small portion of it is picked up by a receiving antenna

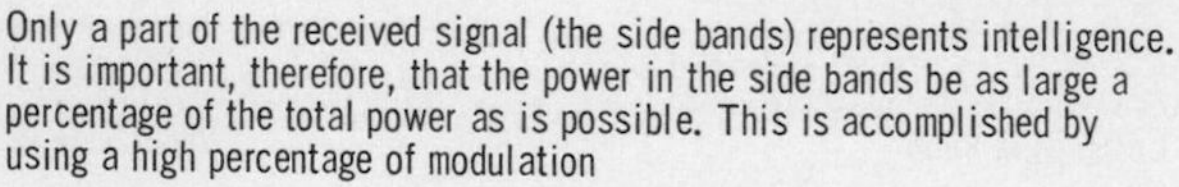

Only a part of the received signal (the side bands) represents intelligence. It is important, therefore, that the power in the side bands be as large a percentage of the total power as is possible. This is accomplished by using a high percentage of modulation

The power in an amplitude-modulated signal is *divided* among the carrier and the side bands. The relative amounts of power in the carrier and the side bands depends on the percentage of modulation. But in all cases, the two side bands generally contain *equal power*, and their combined power is less than that contained in the carrier. At 100-percent modulation, the total side-band power is one-half the carrier power. Thus, the side bands together hold 33.3 percent of the total signal power, or each side band has 16.7 percent of the total power. These percentages drop sharply when the modulation is less than 100 percent.

Since, as you know, the side bands contain the intelligence, the power in the side bands represents *signal power*. It is important, therefore, to have as much power as possible in the side bands; and this is why a high percentage of modulation is normally desirable.

summary

☐ For each modulating frequency, two side-band frequencies are produced: one above the carrier frequency and one below it. ☐ The upper side-band frequencies represent the sum of the individual modulating frequencies and the carrier frequency. ☐ The lower side-band frequencies represent the difference between the modulating frequencies and the carrier frequency.

☐ The bandwidth of an amplitude-modulated signal is the frequency range between the lowest, lower side-band frequency and the highest, upper side-band frequency. The bandwidth is always equal to twice the highest modulating frequency. ☐ Since bandwidth is determined by the frequencies in the modulating signal, all circuits that process a signal must accommodate the entire bandwidth.

☐ The intelligence in an amplitude-modulated signal is carried in the side bands. In effect, the unmodulated carrier merely serves as a means of increasing the low-frequency intelligence of the modulating signal to a higher frequency; namely, the side-band frequencies. ☐ Both side bands of an amplitude-modulated signal carry the identical intelligence. Therefore, to recover the intelligence, all that is required is the unmodulated carrier and one of the side bands. ☐ Power in an amplitude-modulated signal is divided among the carrier and the two side bands, with each band containing equal power. ☐ Since the side bands contain the signal intelligence, it is important to have as much power as possible in the side bands, and this is accomplished by using a percentage of modulation close to 100 percent.

review questions

1. If a 1-kHz tone amplitude modulates a 1-MHz carrier, what are the side-band frequencies?
2. What is the bandwidth of the modulated signal in Question 1?
3. What do the side bands carry?
4. What effect does the carrier frequency have on bandwidth?
5. Is all the power in an AM signal carried in the side bands?
6. If one side band is removed from an AM signal, will the signal power be affected?
7. If one side band is removed from an AM signal, will the signal intelligence be affected?
8. A modulating signal containing three frequencies will produce how many side-band frequencies?
9. If the modulating frequencies in Question 8 are 2, 4, and 6 kHz, what is the bandwidth of the modulated signal?
10. How does percentage of modulation affect side-band power?

single side-band modulation

You have seen that both side bands of an amplitude-modulated signal contain all of the intelligence being transmitted, So, to recover the intelligence, all that is required is one side band and the carrier. If one of the side bands, therefore, was removed from the modulated carrier immediately after modulation, it would have no harmful effects on the transmission of the intelligence. The intelligence would be transmitted in the other side band, and the unmodulated carrier would accompany it for later use in converting the intelligence to its original lower frequency.

This technique of removing one side band from an amplitude-modulated signal is called *single side-band modulation* (SSB), and has certain advantages over conventional transmission in which both side bands are used. The most important advantage to be gained by eliminating one side band is that the bandwidth of the signal is cut in half.

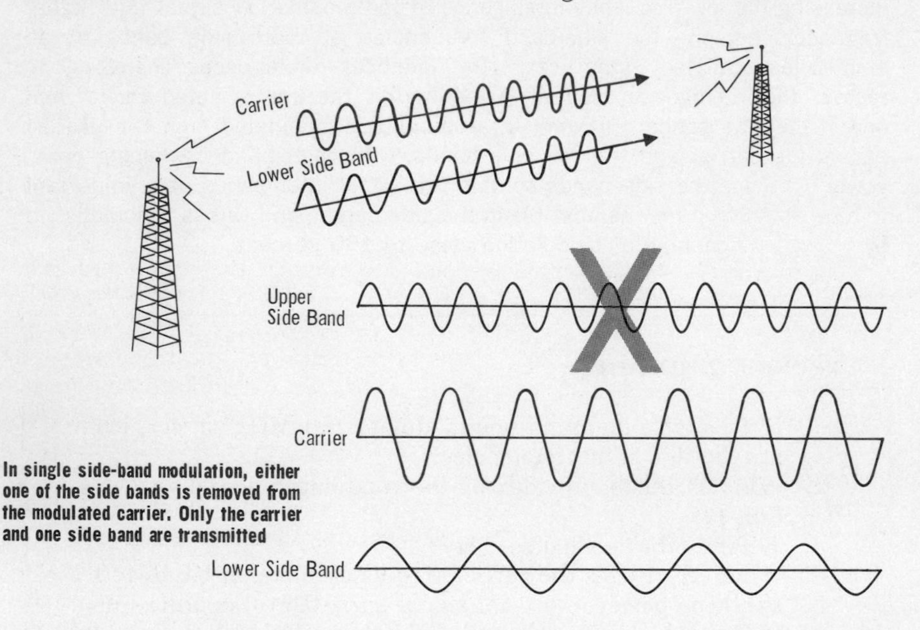

In single side-band modulation, either one of the side bands is removed from the modulated carrier. Only the carrier and one side band are transmitted

You recall that with both side bands, half of the bandwidth is above the carrier frequency and half below. But both halves represent the same intelligence, only at different frequencies. So by eliminating one side band, the range of frequencies that carry the intelligence is cut in half. This reduction in bandwidth improves reception of the signal by receiving equipment and circuits, since the narrower the bandwidth, the less atmospheric noise, or static, that will enter the receiving circuits with the signal. Also, if each carrier uses less bandwidth in a given range, more carrier signals can be sent, or there will be less interference between different carrier signals.

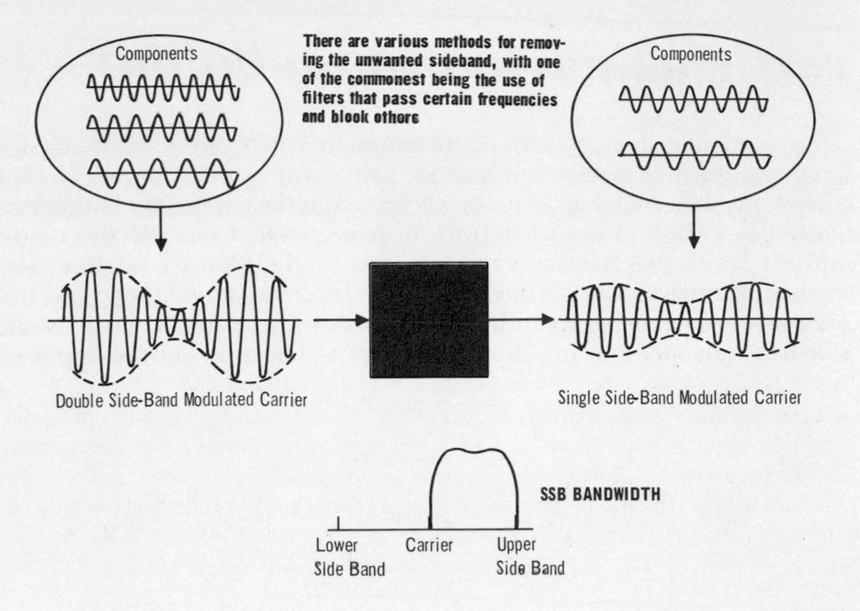

single side-band modulation (cont.)

The filters or other components used to remove the unwanted side band from a single side-band signal are not perfect devices. There is no filter that will completely eliminate, or block, say a frequency of 100 kHz (100,000 Hz), and at the same time pass a frequency of 100,001 or 99,999 Hz without any reduction, or attenuation, in amplitude. Instead, *practical* filters used to eliminate an unwanted side band completely block certain frequencies, attentuate others, and pass all remaining frequencies.

In terms of a frequency spectrum, the attenuated frequencies are on either side of and close to the blocked frequencies, while the frequencies passed with no attenuation are relatively far from the blocked frequencies. As a result of these filter characteristics, it is difficult to remove a side band completely when the modulating signal that produces the side bands contains *low frequencies*. This is because low modulating frequencies produce pairs of side-band frequencies that are *close* to each other, as well as to the carrier. For example, the side-band frequencies produced by a modulating frequency of 50 Hz are only 100 Hz apart and 50 Hz from the carrier; whereas a relatively high modulating frequency of say 2000 Hz will produce side-band frequencies that are 4000 Hz apart and 2000 Hz from the carrier. If an attempt is made, therefore, to completely remove a side band that contains frequencies very close to the carrier, both the carrier and some frequencies of the other side band will be attenuated, and this is undesirable.

vestigial side-band modulation

To overcome this problem of attenuation when low modulating frequencies are involved, the unwanted side band *is not completely eliminated.* Instead, frequencies relatively far from the carrier are eliminated, while those close to the carrier are only *attenuated,* so that the carrier and the other side band are not affected by the filtering process. This method of modulation is called *vestigial side-band modulation,* and has, although to a lesser degree, the narrow bandwidth advantages of single side-band modulation. In effect, only part of one side band is removed.

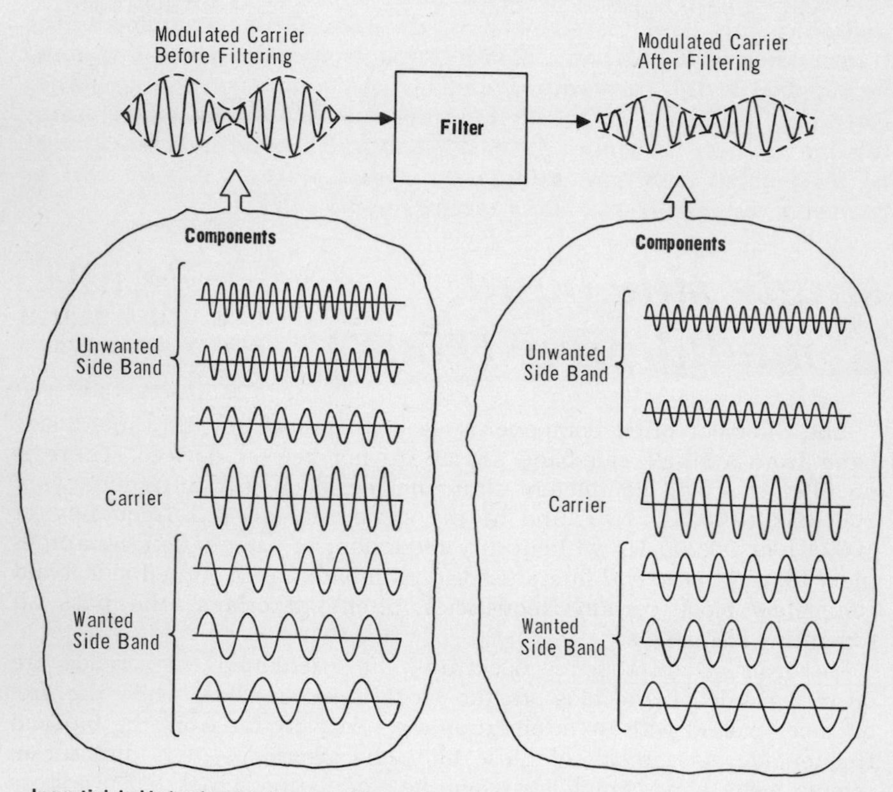

In vestigial side-band modulation, one side band is not completely eliminated as it is in single side-band modulation

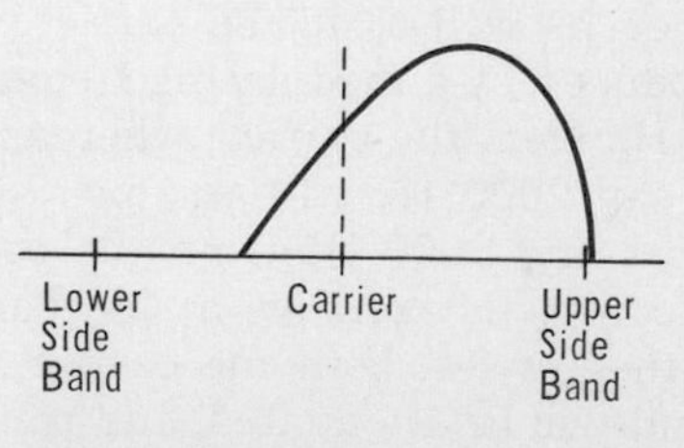

Instead, only some of the frequencies of the unwanted side band are eliminated; those frequencies closest to the carrier are merely attenuated. Sometimes the carrier is also attenuated

suppressed carrier modulation

When you think in terms of side bands, the purpose of the carrier is to convert audio-frequency (a-f) intelligence to radio-frequency (r-f) intelligence before transmission; and after transmission, to reconvert the r-f intelligence to its original audio form. Thus, the side bands (either one, or both) are transmitted to send the intelligence from one place to another. The carrier, although it contains no intelligence, is also transmitted just so it is *available* for demodulation at the receiving end.

Since the carrier contains much more power than the side bands, and since this power is supplied by the transmitting equipment, the transmission of the carrier is a *large percentage* of the power that must be supplied at the transmitting end. A considerable *savings* in power can be accomplished, therefore, by eliminating the carrier before transmission and just transmitting one side band. Of course, the carrier must be regenerated and *reinserted in the signal* at the receiving end to reconvert the intelligence to its original audio form.

Lower Side Band

CARRIER REPRODUCED
AT RECEIVING END
FOR DEMODULATION

SSB SUPPRESSED
CARRIER BANDWIDTH

UPPER SIDE BAND

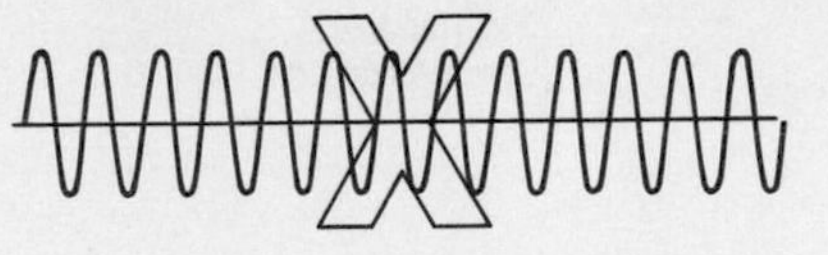

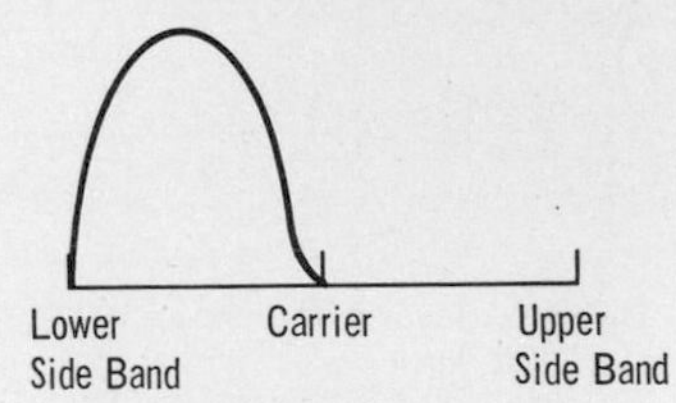

CARRIER

LOWER SIDE BAND

In single side-band suppressed carrier modulation, the carrier and one side band are eliminated from the signal and only one side band is transmitted. At the receiving end, the carrier is regenerated for use in the demodulation process

suppressed carrier modulation (cont.)

The carrier can be eliminated before transmission because, since it is a constant-frequency, constant-amplitude wave, it is easily generated, and at a much lower powel level than that of a transmitted carrier. Actually, the carrier generated at the receiving end might only require a power level of a few watts, whereas transmitted carriers normally have power levels of kilowatts.

The elimination of the carrier, as well as one side band, prior to transmission is called *single side-band modulation with suppressed carrier,* or just *single side-band suppressed carrier modulation.* To some people and in some literature, the term "single side-band modulation" automatically means with a suppressed carrier. To avoid confusion, care should be taken in using the term.

In double side-band reduced carrier modulation, both side bands and the carrier are transmitted

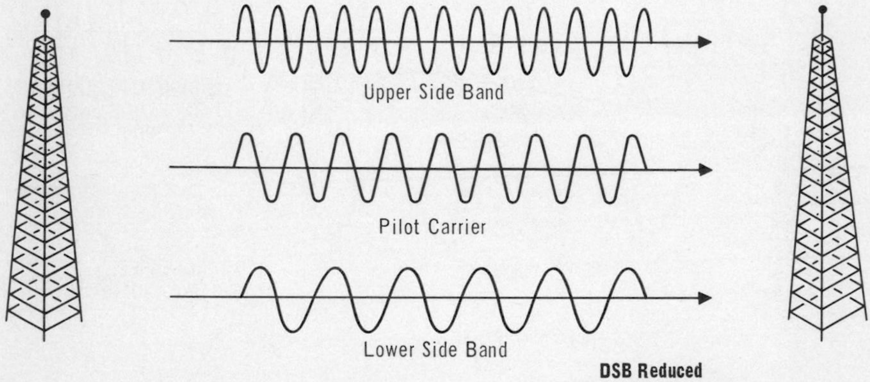

Upper Side Band

Pilot Carrier

Lower Side Band

DSB Reduced Carrier Bandwidth

The power level of the carrier is greatly reduced to a value much less than that of the side bands. The low-level carrier is called the pilot carrier

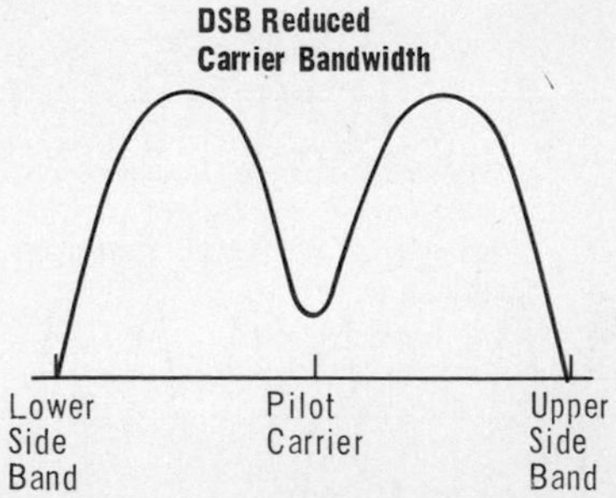

Lower Side Band

Pilot Carrier

Upper Side Band

With the same principles of single side-band suppressed carrier modulation, and for the same reasons, it is also possible to eliminate *only* the carrier, and transmit both side bands. This is *double side-band* (DSB) *suppressed carrier modulation,* and has certain disadvantages when compared to SSB suppressed carrier modulation.

suppressed carrier modulation (cont.)

The main disadvantage of DSB suppressed carrier modulation is the difficulties involved in the demodulation process. To achieve satisfactory demodulation, the phase of the regenerated carrier at the receiving end must be the *same*, or nearly the same, as the phase of the carrier that produced the side bands. If the *difference in phase* is too large, severe distortion of the signal intelligence results. This problem does not exist to as large a degree for SSB suppressed carrier modulation.

To ensure the correct phase relationship between the carrier used for modulation, the carrier at the transmitting end (modulation carrier) is not completely eliminated when DSB modulation is employed. Instead, the carrier is transmitted along with the side bands, but at a greatly *reduced* power level. Such a reduced carrier is called a *pilot carrier*, and is used at the receiving end of the system to maintain the correct phase of the regenerated carrier.

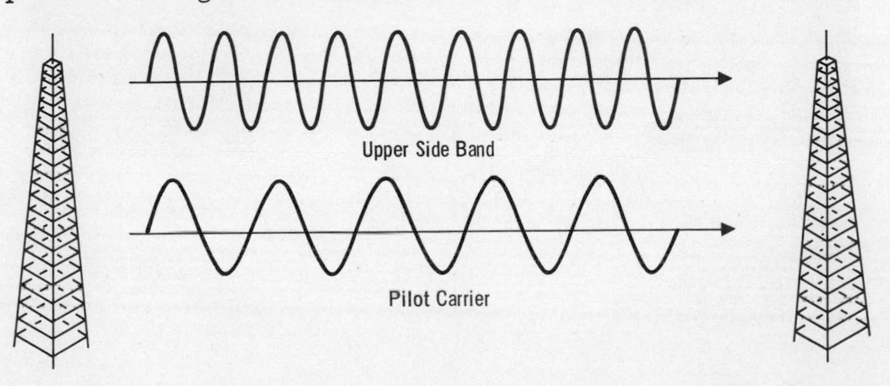

Single side-band reduced carrier modulation is the same as DSB reduced carrier, except that only one side band is transmitted with the pilot carrier

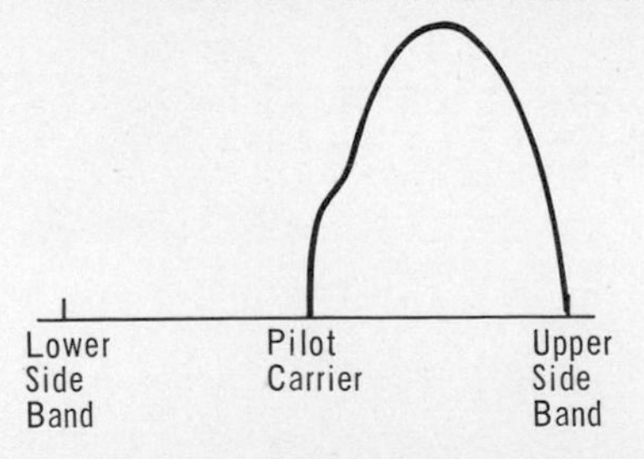

A pilot carrier is often used with SSB modulation, especially when music is the intelligence transmitted. When both side bands are transmitted with a pilot carrier, it is DSB reduced carrier modulation; and when only one side band and a pilot carrier are used, it is SSB reduced carrier modulation.

types of amplitude modulation

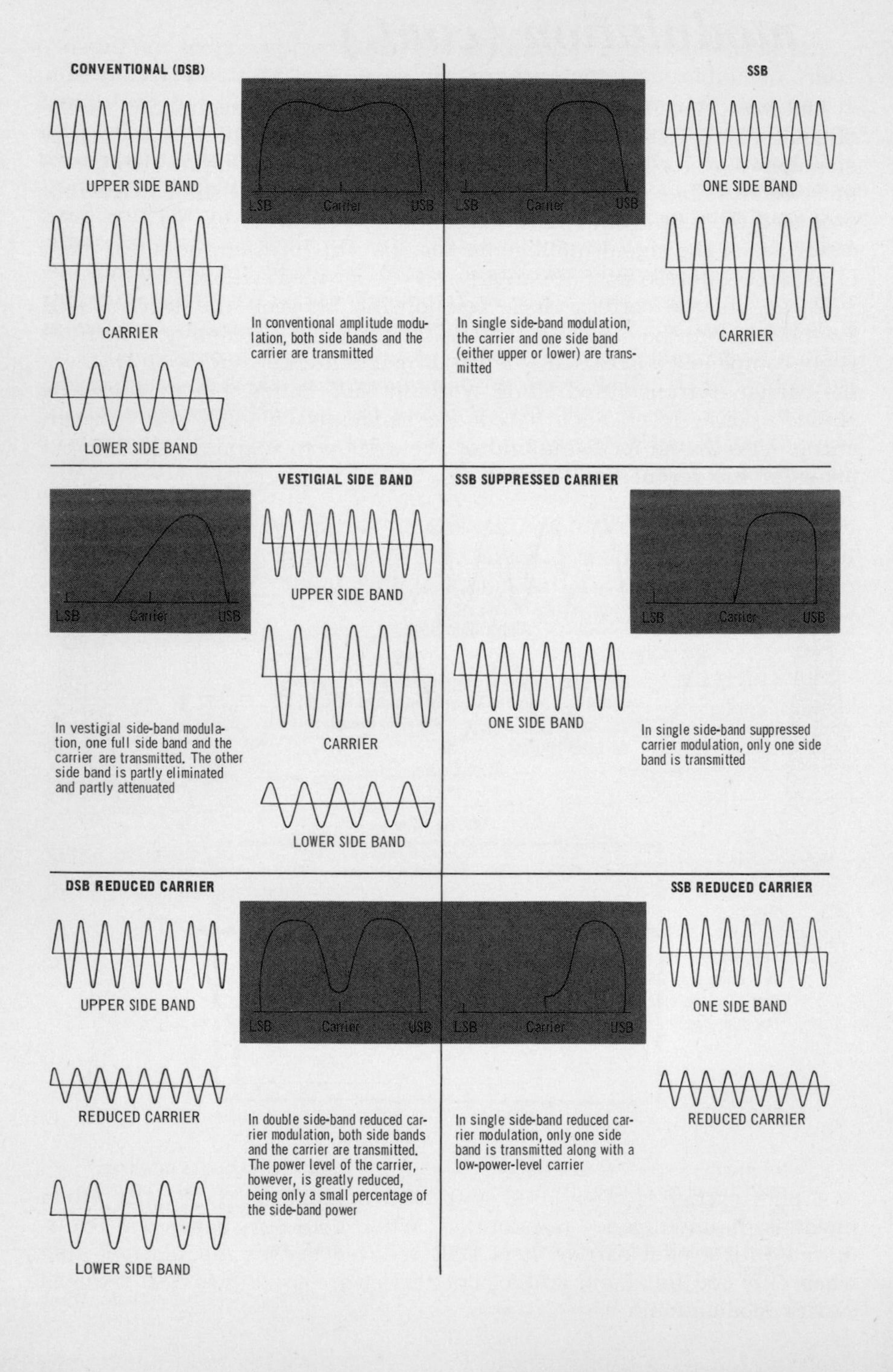

AM demodulation or detection

AM *demodulation* is the process of *recovering* the signal intelligence from an amplitude-modulated carrier wave. It is also called *detection.* It can best be described from the standpoint of the overall modulated carrier wave rather than from its components, which you know are side bands and an unmodulated carrier. If the carrier was suppressed or reduced before transmission, a carrier of the proper phase, frequency, and amplitude must first be generated and combined with the side band before the signal can be demodulated. This is called *carrier reinsertion.*

When an amplitude-modulated signal is ready for demodulation, therefore, it consists of a carrier frequency whose peak-to-peak amplitude varies in accordance with the intelligence being carried. The demodulation or detection process consists of sending such a modulated wave through a circuit called a *detector* or *demodulator*, which does two operations to the wave: it first cuts off either the top or bottom half of the wave (this is called *rectification*); and then the detector removes the r-f portion of the remaining half of the wave, but leaves a signal that follows the *envelope* of that half of the wave. In effect, therefore, the detector eliminates all of the modulated carrier wave, except one-half of the envelope. And since the variations of both halves of the envelope represent the intelligence, reducing the wave to one-half of the envelope completes the demodulation process.

AM demodulation is accomplished by removing everything from the modulated signal except one-half of the envelope, which represents the intelligence

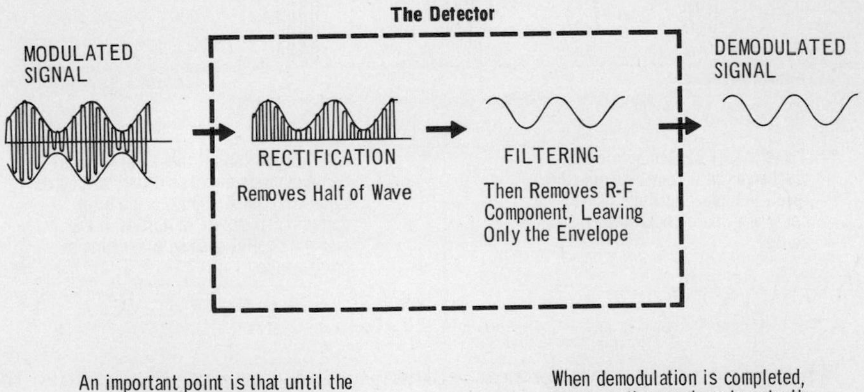

An important point is that until the completion of demodulation, the envelope of the modulated signal is not a physical quantity – it just shows amplitude variations of the carrier

When demodulation is completed, however, the envelope is actually a varying voltage or current that corresponds to the original intelligence

disadvantages

As a means of carrying intelligence, amplitude modulation has many advantages. However, it also has some disadvantages that under certain conditions limit its usefulness and make other forms of modulation more desirable. The major disadvantage of amplitude modulation is that it is easily affected by atmospheric noise (static), other electronic signals having similar frequencies, and interference from such electrical equipment as motors and generators. All of these tend to *amplitude modulate* a carrier in the same way as does its own modulating signal. And as a result, they become a part of the modulated signal, and remain with it right through the demodulation process. After demodulation, they appear as *noise* or *distortion,* which, if severe enough, can completely mask the intelligence and make the demodulated signal worthless. Even if they are not severe enough to diminish intelligibility, they can be extremely annoying.

Probably the most common type of interference that affects amplitude-modulated signals is atmospheric static, which is especially severe during violent thunderstorms

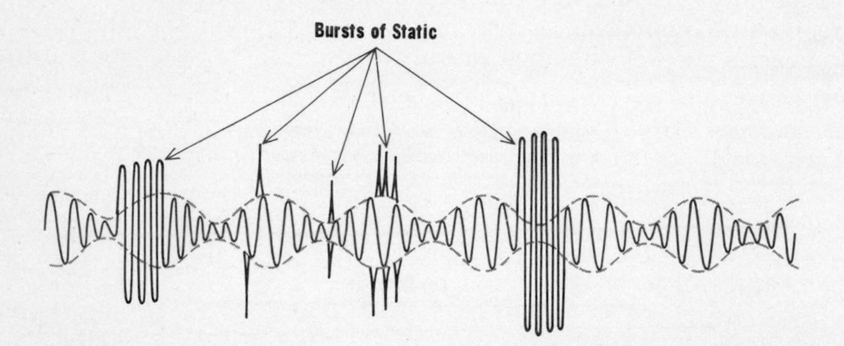

The static is caused by electrical discharges in the atmosphere which result in large-amplitude pulses being amplitude modulated on the carrier

After demodulation, these pulses appear as loud bursts of noise if they are on a voice-modulated signal, or as brief periods of complete distortion if the signal is carrying some other form of intelligence

The only way to prevent or eliminate interference that tends to amplitude modulate a carrier is to place the intelligence on the carrier in some way other than by amplitude variations. In other words, use some type of modulation other than amplitude modulation. One such type of modulation that has good *interference-resistant* properties is *frequency modulation.*

summary

☐ Single side-band modulation (SSB) is a form of amplitude modulation in which one side band is removed from the signal. ☐ In vestigial side-band modulation, the unwanted side band is not completely removed.

☐ In single side-band suppressed carrier modulation, both carrier and one side band are removed from the signal before transmission; only one of the side bands is transmitted. ☐ SSB modulation with suppressed carrier requires considerably less transmitted power than does a comparable conventional SSB signal. This is because the carrier, which is suppressed, represents a large percentage of the total signal power. ☐ When single side-band suppressed carrier modulation is used, the carrier must be regenerated and reinserted in the signal at the receiving end to reconvert the intelligence to its original form. This adding of the carrier back into the signal is called carrier reinsertion. ☐ To ensure that the reinserted carrier has the required phase, a pilot carrier is sometimes used. This means that instead of being suppressed, the carrier is transmitted at a greatly reduced power level.

☐ Demodulation, or detection, is the process of recovering the signal intelligence from a modulated carrier wave. It is accomplished by a circuit called a detector or demodulator. ☐ The first step in detection is rectification, in which one-half of the modulated carrier wave is removed. Then the r-f portion of the remaining half of the wave is removed. This leaves a signal that follows the envelope of one-half of the modulated wave. ☐ The major disadvantage of amplitude modulation is that it is easily affected by static, other electronic signals, and interference from electrical motors and generators.

review questions

1. How does a *vestigial side-band signal* differ from a single side-band signal?
2. Can single side-band modulation be used when the modulating signal consists of many frequencies?
3. What is the principal advantage of suppressed carrier modulation?
4. A carrier transmitted at a greatly reduced power level is called ____________.
5. What is meant by *carrier reinsertion*?
6. Name three types of single side-band modulation.
7. What is *demodulation*?
8. What is another name for demodulation?
9. What is the major disadvantage of amplitude modulation?
10. What is the major advantage of single side-band modulation?

frequency modulation

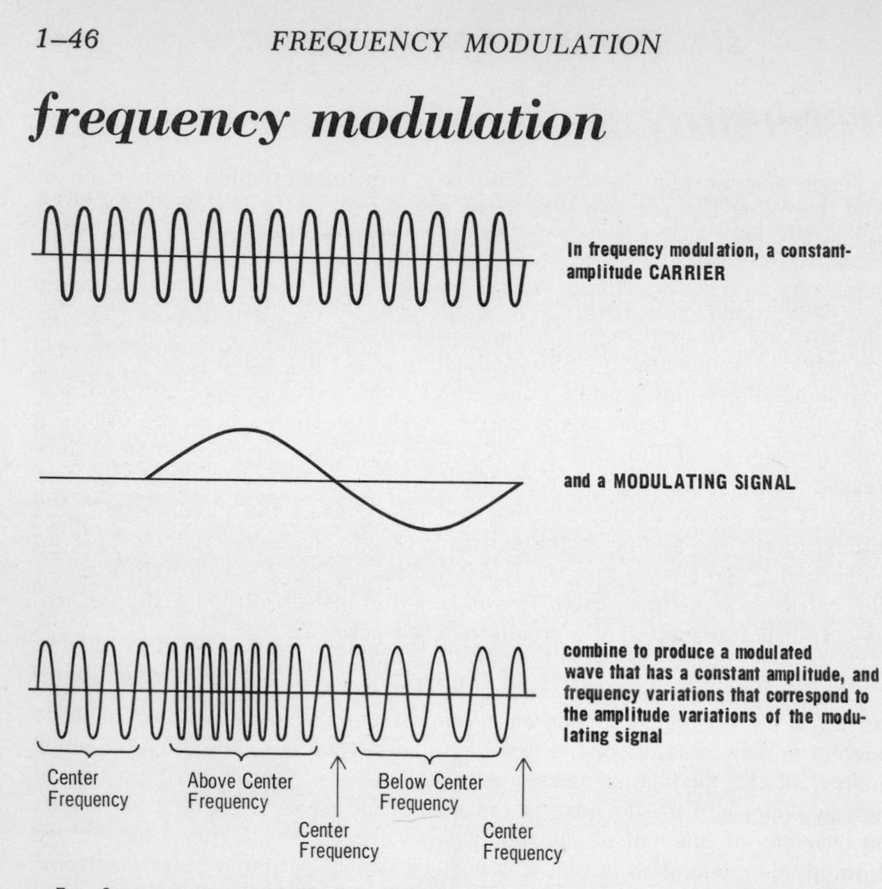

In frequency modulation (FM), an r-f carrier wave is varied in accordance with a modulating signal. However, in *AM*, the *amplitude* of the carrier is changed, whereas in *FM*, it is the *frequency* of the carrier that is varied. When the carrier is frequency modulated, its amplitude is unchanged, while its frequency increases and decreases in accordance with the *amplitude* variations of the modulating signal. The frequency of the unmodulated carrier is called the *center* or *resting* frequency; the carrier fluctuates above and below its center frequency.

A frequency-modulated carrier is at its center frequency when the modulating signal has *zero amplitude*. As the amplitude of the modulating signal increases in the *positive* direction, the frequency of the carrier also *increases*, reaching a maximum when the amplitude of the modulating signal is at its maximum positive value. Then, when the modulating signal decreases in amplitude, the frequency of the carrier also decreases, returning to its center frequency when the modulating signal again reaches zero amplitude.

In the same way, the frequency variations of the carrier follow the *negative* amplitude variations of the modulating signal, except that the carrier frequency *decreases* as the modulating signal becomes more negative, and then increases, reaching its center frequency again when the modulating signal completes its negative half cycle and returns to zero.

frequency generation

You have seen that the frequency of an FM wave varies, or *deviates*, above and below its center frequency according to the amplitude variations of the modulating signal. The total range, or swing from the center frequency to the *lowest frequency*, which corresponds to the *maximum negative amplitude* of the modulating signal, or from the center frequency to the *highest frequency*, which corresponds to the *maximum positive amplitude* of the modulating signal, is called the *maximum frequency deviation* of the carrier.

It is obvious that the greater the amplitude of the modulating signal, the larger is the frequency deviation of the FM carrier. For example, a weak (small-amplitude) modulating signal might cause a carrier having a center frequency of 100 MHz to vary from a low frequency of 99.99 MHz to a high frequency of 100.01 MHz. The maximum frequency deviation would thus be 100.01 MHz minus 100.00 MHz, or ± 10 kHz. A strong signal, on the other hand, might vary the same carrier from 99.95 to 100.05 MHz, for a maximum frequency deviation of ±50 kHz. Thus, the frequency deviation of the carrier indicates the amplitude of the modulating signal.

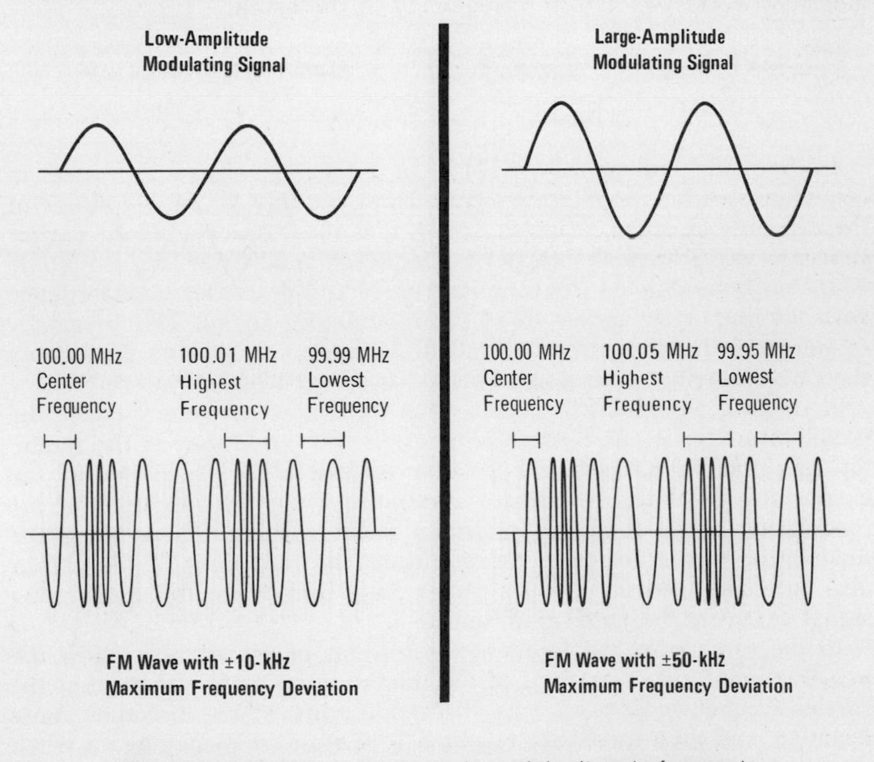

The total frequency swing of an FM wave above or below its center frequency is called the maximum frequency deviation, and is proportional to the amplitude of the modulating signal. The amplitude of the modulated wave is unaffected by the amplitude of the modulating signal

rate of frequency deviation

The amplitude of the modulating signal determines the maximum frequency deviation of the modulated carrier. Each time the modulating signal goes through one full cycle, assuming it is sinusoidal, the carrier also goes through *one full frequency deviation.* This consists of starting at the center frequency, increasing to its maximum frequency, then decreasing to its minimum frequency, passing through the center frequency on the way, and finally increasing back to the center frequency. Thus, if the modulating signal has a frequency of 10 kHz, the carrier goes through 10,000 full frequency deviations every second. In other words, the rate of frequency deviation of the carrier, or the rate at which the carrier deviates, is determined by the frequency of the modulating signal. How far the carrier deviates, though, is still only affected by the amplitude of the modulating signal.

You can see now that the intelligence carried by an FM wave is represented by the maximum frequency deviation of the carrier, as well as by the rate at which the carrier goes through the maximum deviation. In the case of audio intelligence, the amount of carrier deviation corresponds to the amplitude or loudness of the sound, while the rate of deviation corresponds to the frequency of the sound.

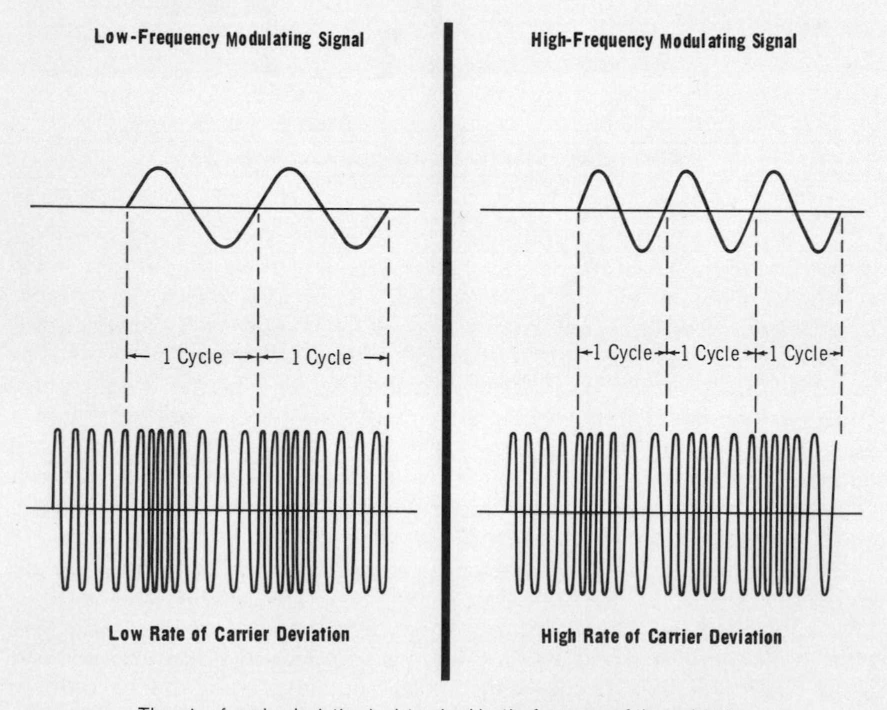

The rate of carrier deviation is determined by the frequency of the modulating signal. Low-frequency intelligence, therefore, causes slow deviation of the carrier frequency, while high-frequency intelligence causes the carrier to deviate more rapidly

generation of side bands

During the process of frequency modulation, just as during amplitude modulation, new frequencies, called side-band frequencies, are produced above and below the unmodulated carrier frequency. These side-band frequencies contain the signal intelligence, as in amplitude modulation, and combine with the unmodulated carrier to produce the modulated carrier previously described.

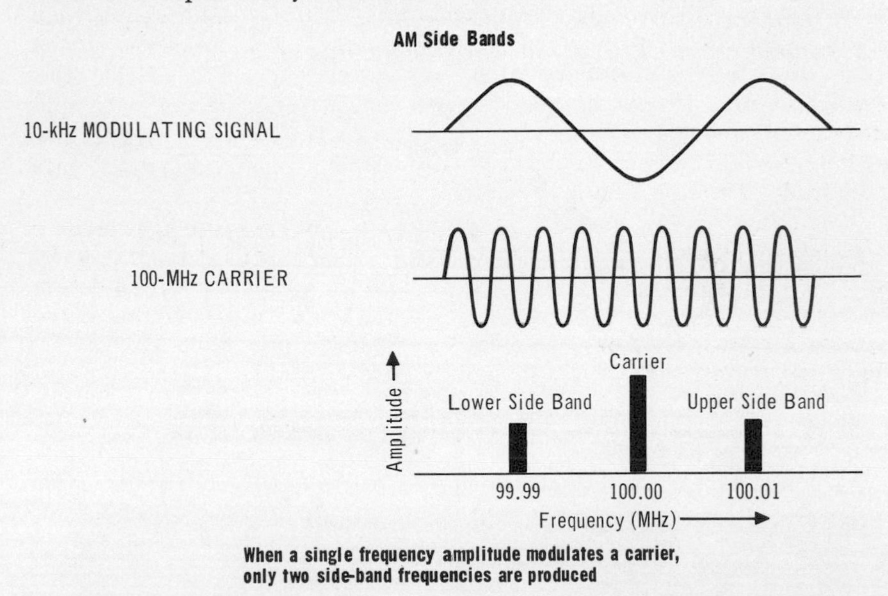

When a single frequency amplitude modulates a carrier, only two side-band frequencies are produced

A significant difference between AM and FM side-band frequencies is the *number produced.* If you recall, in amplitude modulation, two side-band frequencies are produced for every modulating frequency. One of these side-band frequencies is equal to the *sum* of the modulating and carrier frequencies, and is *above* the carrier frequency. The other is equal to the *difference* between the modulating and carrier frequencies, and is *below* the carrier. In FM, each modulating frequency produces a similar pair of sum and difference side-band frequencies. However, in addition to the basic pair, a theoretically infinite number of additional side-band frequencies are produced. These additional frequencies are equal to *whole number multiples* of the basic pair.

For example, if a 1-MHz carrier is frequency modulated by a single frequency of 10 kHz, the basic pair of side-band frequencies will be 1010 kHz and 990 kHz. The additional frequencies will be at 1020 and 980 kHz, 1030 and 970 kHz, 1040 and 960 kHz, and so on. Although theoretically these side-band frequencies extend outward from the carrier indefinitely, only a limited number of them contain *sufficient power* to be significant. Even so, this limited number is still always far greater than the number produced by comparable amplitude modulation.

generation of side bands (cont.)

The number of significant side-band frequencies produced in any particular case depends on the amplitude and frequency of the modulating signal. The larger the amplitude or the lower the frequency of the modulating signal, the greater is the number of significant side-band frequencies. The exact number can be found using a ratio called the *modulation index*. You will learn about this later.

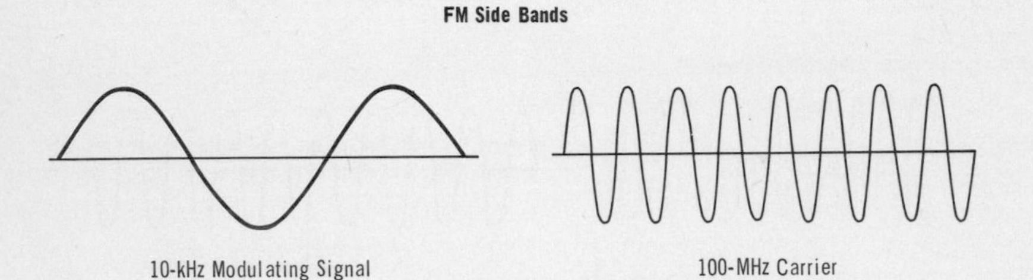

When a single frequency frequency modulates a carrier, many side-band frequencies are produced. Each of these side-band frequencies is separated from adjacent ones by the modulating frequency

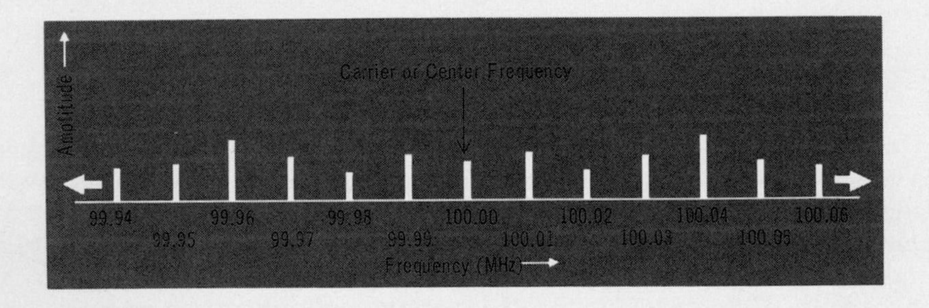

You recall that in amplitude modulation, the amplitudes of the side-band frequencies, or the power contained in them, were independent of the amplitude of the unmodulated carrier, and depended only on the amplitude, or power, of the modulating signal. In frequency modulation, the situation is different. The side bands derive their power from the carrier, which means that the unmodulated carrier component of an FM wave has less power, or smaller amplitude, *after* modulation than it does before. The amount of power removed from the carrier and placed in the side bands depends on the modulating frequencies and the maximum deviation of the carrier. It is possible, under certain conditions, for the carrier power to be *zero*, with all of the power in the side bands. This, of course, is desirable, since the carrier itself contains no intelligence.

generation of side bands (cont.)

The amplitudes of the individual side-band frequencies depend on the *modulation index,* which is described later. The pattern of the individual amplitudes is highly *irregular.* There is no continuous increase in amplitude as frequencies go further from the carrier, nor is there a continuous decrease. However, in all cases, there is a point relatively distant from the carrier where the amplitudes of the side-band frequencies drop below *1 percent* of the amplitude of the unmodulated carrier. Past this point, the side bands are insignificant, and can be ignored.

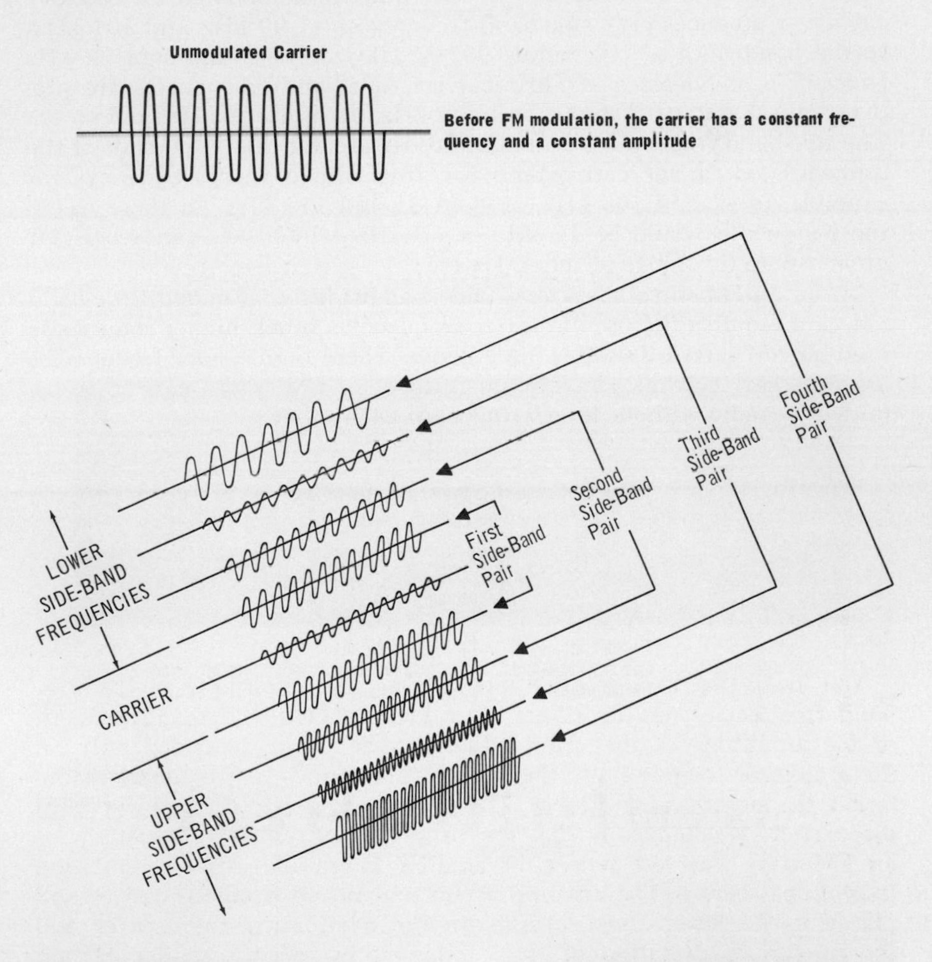

After modulation, the carrier component of the FM wave still has the same constant frequency, but its amplitude has been reduced as a result of the power taken from it by the side bands

The side-band frequencies also have constant amplitudes, which when combined with the carrier component produce the modulated wave with its constant amplitude but varying frequency

FM bandwidth

In AM, you learned that the term *bandwidth* meant the entire range of frequencies in a modulated wave. Because of the many side-band frequencies contained in an FM wave, bandwidth when applied to FM is more restrictive: It includes only the *significant frequencies*. The bandwidth of an FM wave is the frequency range between the *extreme upper* and *extreme lower* side-band frequencies whose amplitudes are 1 percent or more of the unmodulated carrier amplitude. Since these extreme side-band frequencies are multiples of the modulating frequencies, you can see that the bandwidth of an FM wave can be many times greater than that of an AM wave.

For example, if a frequency of 1 kHz amplitude modulates a 100 kHz carrier, it produces only side-band frequencies at 99 kHz and 101 kHz; so the bandwidth is 101 minus 99, or 2 kHz. But if the same 1 kHz *frequency modulates* a 100-kHz carrier, side-band frequencies are produced at 99 and 101 kHz, 98 and 102 kHz, 97 and 103 kHz, and so on. The side-band frequencies with amplitudes greater than 1 percent of the unmodulated carrier can extend far from the carrier frequency; for example, to 92 and 108 kHz, or even 81 and 119 kHz. In these cases, the bandwidth would be 16 kHz or 38 kHz, which you can see is far greater than the 2 kHz of the AM wave.

When an FM wave has a very wide bandwidth, it is called *wide-band FM*, and requires the use of carrier frequencies much higher than those used for AM carrying similar intelligence. These high carrier frequencies are necessary so that a maximum number of FM waves can be transmitted by radio without interfering with each other.

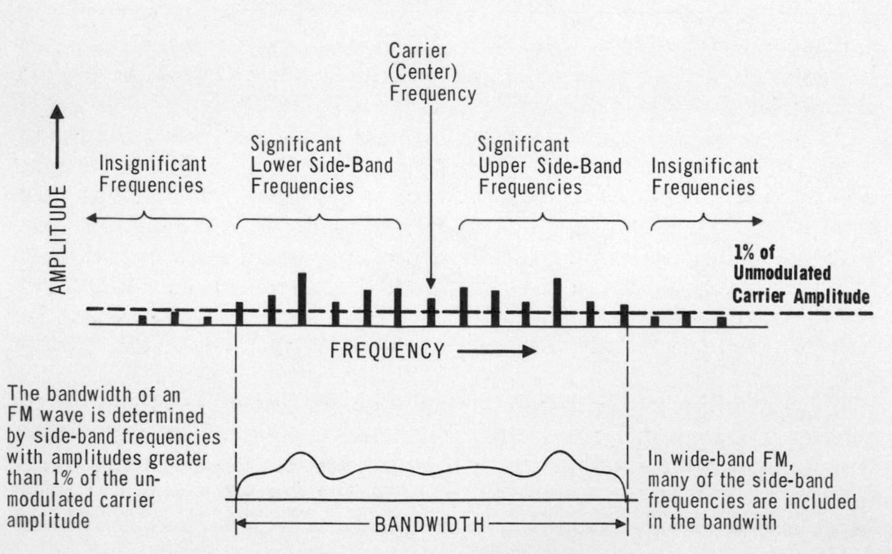

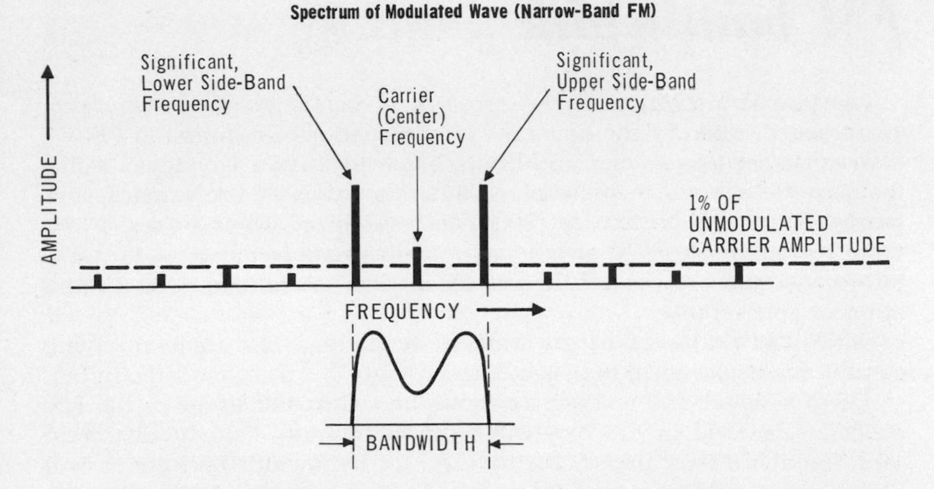

In narrow-band FM, only the two basic side-band frequencies have significant amplitudes. The bandwidth of such a wave, therefore, is the same as an AM wave

FM bandwidth (cont.)

It is possible, by limiting the maximum deviation of the FM carrier, to produce FM having the same bandwidth as an AM wave carrying the same intelligence. This is called *narrow-band FM*. While this process causes some distortion of the intelligence, it allows carrier frequencies to be used that are lower than some required for wide-band FM.

You should understand at this point that the bandwidth of a modulated wave is important for two reasons: first, it determines how much *space* or room in the radio-frequency spectrum the wave will occupy; and second, it determines the range of frequencies over which the electronic circuits used to receive and process the wave must be capable of operating.

As far as the radio-frequency spectrum is concerned, all of the modulated waves transmitted by radio in any one geographic area must occupy *different* places in the spectrum or else they will interfere with each other. For example, the lower frequencies of a modulated wave with a 20-kHz carrier and an 8-kHz bandwidth would overlap and interfere with the upper frequencies of a 14-kHz carrier with an 8-kHz bandwidth. You can see that interference between radio waves can be avoided either by *reducing* bandwidths or by moving carrier frequencies *farther apart.*

If bandwidths are made too narrow, though, distortion of the intelligence carried by the wave will result, since many of the side bands, which contain the intelligence, will be eliminated. On the other hand, if carrier frequencies are too far apart, a very limited number of radio waves would *completely fill* the radio spectrum.

FM bandwidth (cont.)

The practical solution to the problem of bandwidth allocation is a compromise, whereby the Federal Communications Commission (FCC) assigns carrier frequencies and limits bandwidths to a frequency range that is wide enough to prevent extreme distortion of intelligence, and narrow enough to prevent interference between adjacent waves in the radio spectrum. The FCC also licenses and assigns frequencies to transmitter operations to control the use of airwaves so that there is a minimum of interference.

With bandwidths set by government regulation, electronic receiving circuits can be designed to respond accordingly. To receive all the intelligence in a signal, the receiving circuits must respond to all of the frequencies included in the bandwidth. If the circuits cannot, distortion over and above that already set by the FCC bandwidth limitations will occur. You should be aware here that the distortion of the intelligence caused by the FCC bandwidth limitations is insignificant, since only the extreme side-band frequencies are eliminated, and these contribute little to the overall intelligence.

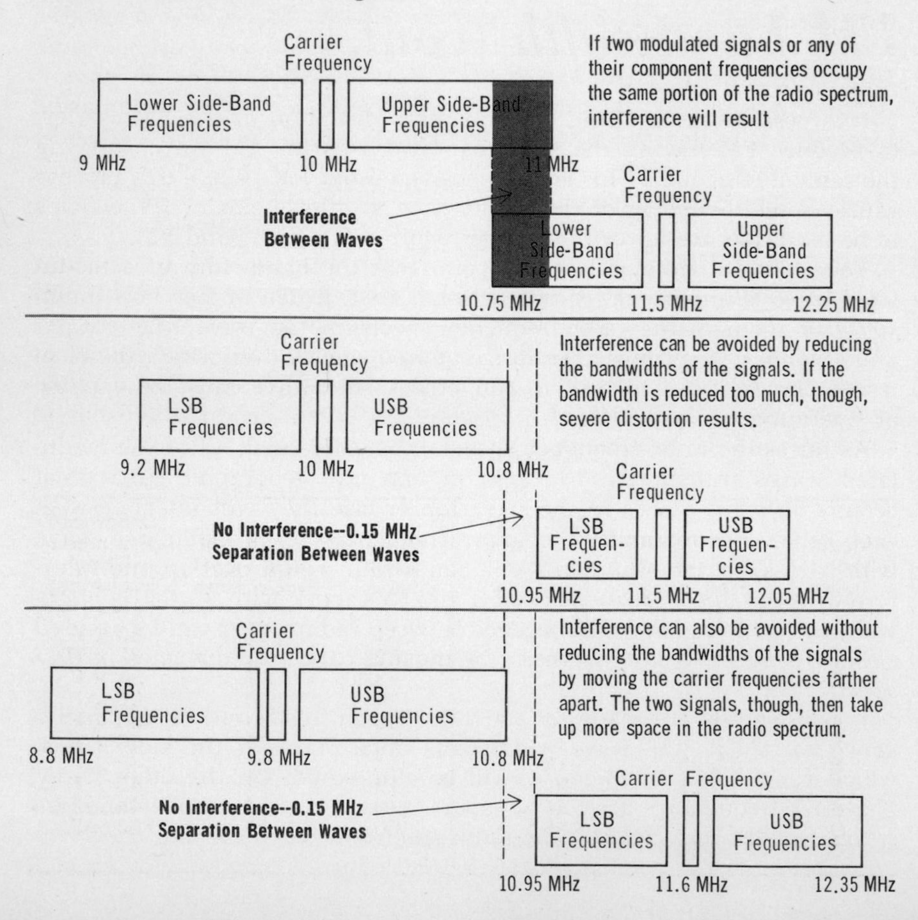

the modulation index

Since the bandwidth is the range between the upper and lower *significant* side-band frequencies, you may think that the more significant side-band frequencies there are, the wider is the bandwidth. This is not always the case. Remember that with a single modulating frequency, the side-band frequencies are separated by a space equal to the modulating frequency. Thus, the side-band frequencies produced by high modulating frequencies are much farther apart than are those produced by low modulating frequencies.

It is possible to have a relatively wide bandwidth when the side-band frequencies are far apart even if only a few of them have significant amplitudes. Similarly, closely spaced side-band frequencies produced by a low modulating frequency can result in a relatively narrow bandpass even though many of them have significant amplitudes. The situation is further complicated by the fact that the number of side-band frequencies that have significant amplitudes depends on the *maximum frequency deviation* of the FM carrier, which in turn, as you know, depends on the amplitude of the modulating signal. You can see, then, that the bandwidth of an FM wave is determined, in a complicated way, by the frequency (or highest frequency) of the modulating signal, and the maximum deviation of the carrier caused by that frequency.

The ratio of the maximum carrier deviation to the modulating frequency is called the *modulation index:*

$$\text{Modulation index} = \frac{\text{maximum carrier deviation}}{\text{maximum modulating frequency}}$$

Using a special form of mathematics, tables like the one below show the relationship between the modulation index, and the bandwidth and number of significant side-band frequencies of an FM wave. As an example of how to use the table, consider a modulating frequency of 1 kHz that causes a maximum carrier deviation of 7 kHz. The modulation index is 7 kHz/1 kHz, or 7. So there are 11 significant side bands in the wave, and the bandwidth is 22 × 1 kHz, or 22 kHz.

Modulation Index	Number of Side-Band Frequencies	Bandwidth (F = modulating frequency)	Modulation Index	Number of Side-Band Frequencies	Bandwidth (F = modulating frequency)
0.5	2	4 × F	11	15	30 × F
1	3	6 × F	12	16	32 × F
2	4	8 × F	13	17	34 × F
3	6	12 × F	14	18	36 × F
4	7	14 × F	15	19	38 × F
5	8	16 × F	16	20	40 × F
6	9	18 × F	17	21	42 × F
7	11	22 × F	18	23	46 × F
8	12	24 × F	19	24	48 × F
9	13	26 × F	20	25	50 × F
10	14	28 × F			

percentage of modulation

In AM, you recall, the percentage of modulation expressed the degree to which the modulating signal caused the peak-to-peak amplitude of the carrier to vary. Thus, at 100 percent amplitude modulation, the peak-to-peak *amplitude* of the carrier varies between zero and twice its unmodulated value. If the same type of system was used to express the percentage of modulation of FM waves, 100 percent modulation would mean that the carrier *frequency* varied between zero and twice its unmodulated value. This is entirely impractical; so instead, the percentage of modulation for FM is defined in terms of the maximum frequency *deviation* that can be produced by the electronic equipment generating the wave.

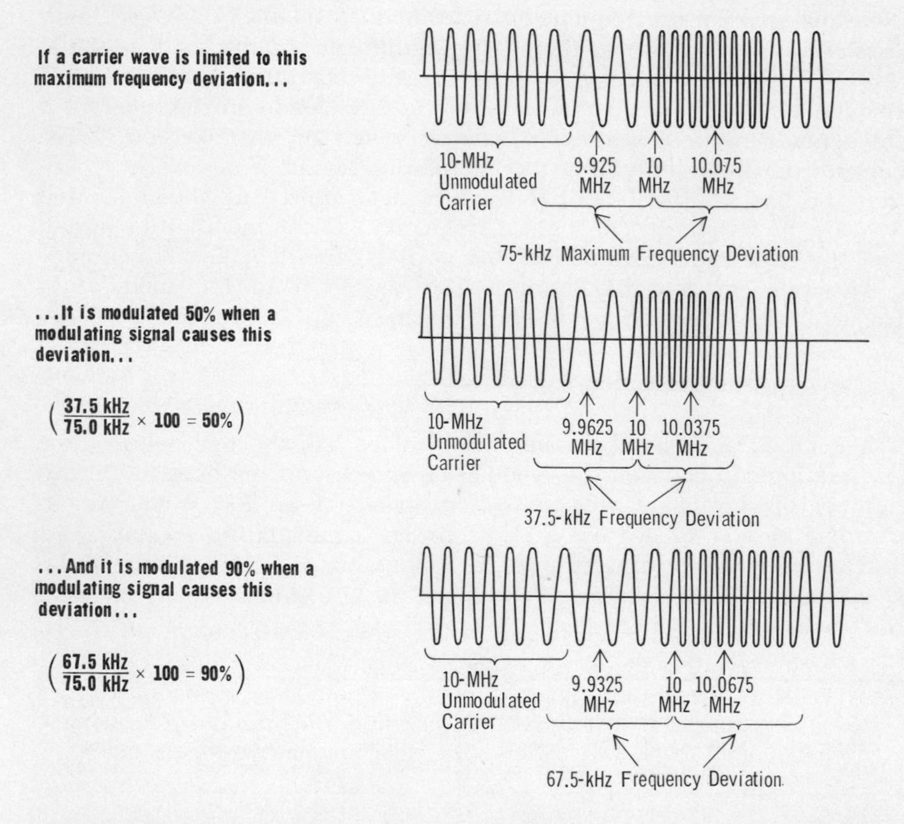

Generally, this maximum frequency deviation is set by FCC regulations, and the percentage of modulation is then the percentage of this maximum deviation that is produced by a modulating signal. For example, the maximum carrier deviation set by the FCC for commercial FM broadcasting is 75 kHz. If a modulating signal causes the full 75-kHz deviation, it has undergone 100 percent modulation. Similarly, a deviation of 37-1/2 kHz represents a modulation of 50 percent.

summary

☐ In a frequency-modulated signal, the frequency of an r-f carrier varies in accordance with the amplitude and frequency of a modulating signal. ☐ The frequency of the unmodulated carrier is called the center, or resting, frequency. When the carrier is frequency modulated, it then fluctuates above and below its center frequency. ☐ The frequency swing from the center frequency to either the highest or lowest modulated frequency is called the frequency deviation. The larger the amplitude of the modulating signal, the greater is the frequency deviation. ☐ The rate of frequency deviation is determined by the frequency of the modulating signal.

☐ Theoretically, each modulating frequency produces an infinite number of side-band frequencies in an FM signal. However, the bandwidth of an FM signal is the frequency range between the extreme upper and lower side-band frequencies whose amplitudes are one percent or more of the unmodulated carrier amplitude. ☐ A wide-band FM signal has a much wider bandwidth than an AM signal that carries the same intelligence. ☐ A narrow-band FM signal has the same bandwidth as an AM signal carrying the same intelligence.

☐ The exact number of significant side-band frequencies in an FM signal can be found by applying a ratio called the modulation index. This is the ratio of the maximum carrier deviation to the modulating frequency. ☐ In frequency modulation, the side bands derive their power from the carrier. Thus, the unmodulated carrier component of the signal has less power after being modulated than it does before.

review questions

1. What is another name for the center *frequency* of an FM carrier?
2. What is meant by the *maximum frequency deviation* of an *FM signal*?
3. How does the amplitude of the modulating signal affect the frequency deviation of an FM signal?
4. How does the frequency of the modulating signal affect the frequency deviation of an FM signal?
5. In an FM signal, which side-band frequencies are significant?
6. What is *narrow-band FM*?
7. What is the disadvantage of narrow-band FM?
8. What is the modulation index used for?
9. Can the power in the unmodulated carrier component of an FM signal ever be zero?
10. Explain the reason for your answer to Question 9.

preemphasis

When a carrier is modulated by a complex signal, the deviation of the carrier depends only on the *amplitudes* of the modulating frequencies, not on the frequencies themselves. In human speech, or in music, *higher frequencies* have *lower amplitudes* than do the lower frequencies. So, in an FM wave carrying speech or music, higher frequencies have smaller deviations than the lower frequencies. If not for noise, this would be no problem. But, when a modulated wave is demodulated, the noise that it "picked up" during its transmission is demodulated as well.

Although the noise is present at all frequencies, the ratio of the noise level to the signal level is greater at the higher (lower-amplitude) frequencies: the signal-to-noise ratio is less at high frequencies. To increase the signal-to-noise ratio of the higher frequencies, certain changes, called *preemphasis,* are made before frequency modulation takes place. These changes consist of emphasizing the amplitudes of the higher modulating frequencies.

Preemphasis distorts the sound signal to some extent, but after transmission and demodulation, the reverse of preemphasis, called *deemphasis,* takes place. The frequencies are then reduced in amplitude to restore the sound signal back to its original form.

It may not seem obvious that these processes increase the signal-to-noise ratio of the higher frequencies, but they do. The important point is that preemphasis increases the level of the high frequencies *before* the noise is encountered, while deemphasis reduces the high frequencies back to their *normal levels,* but at the *same time reduces* the *noise* accompanying these frequencies to *below* its normal level.

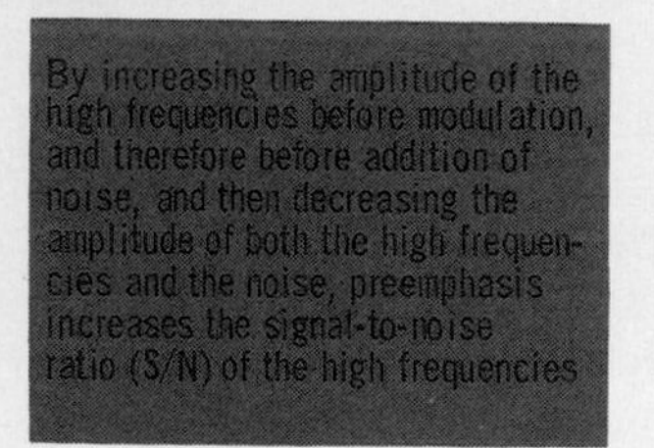
By increasing the amplitude of the high frequencies before modulation, and therefore before addition of noise, and then decreasing the amplitude of both the high frequencies and the noise, preemphasis increases the signal-to-noise ratio (S/N) of the high frequencies

WITHOUT PREEMPHASIS

10-kHz Component of FM Wave (3 Volts)

Noise (1 Volt)

S/N = 3/1 = 3

WITH PREEMPHASIS

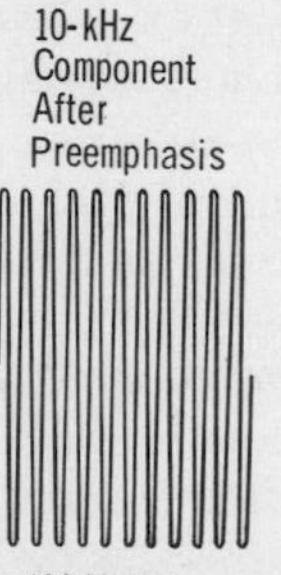

Noise (1 Volt)

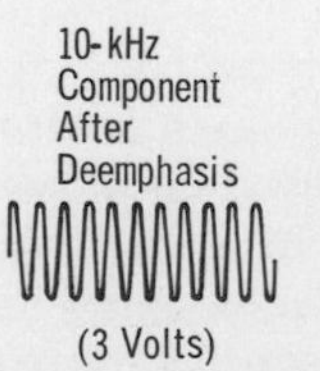

Noise After Deemphasis (0.01 Volt)

S/N = 3/0.01 = 300

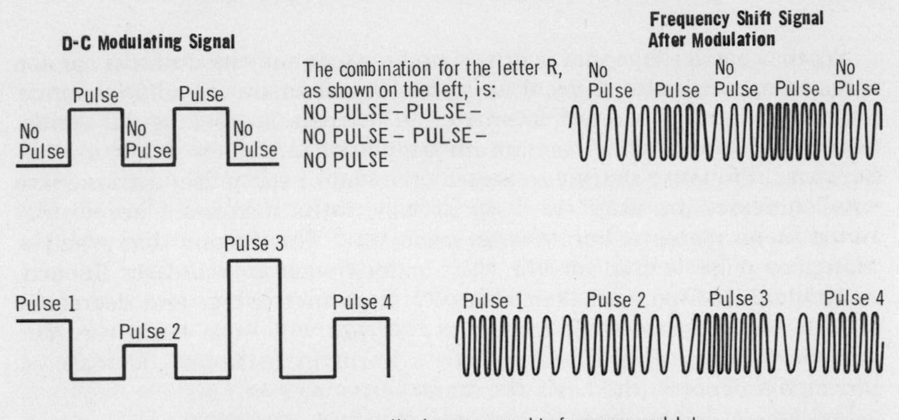

If pulses of various amplitudes were used to frequency modulate a carrier, the modulated signal would have a frequency for each different pulse amplitude, and one for the no-pulse condition

FM with a pulse

Until now, we have considered frequency modulation in which the modulating signal was either continuous, such as a steady tone, or was complex, such as voice or music. Frequently, though, nonsinusoidal pulse-type modulating signals are used with FM. One of the most common types of pulse FM signals is the *frequency shift signal*, which is used with frequency shift keying (FSK), or frequency shift telegraphy. In one method, the modulating signal consists of a series of rectangular d-c pulses, all having *equal* amplitudes. The *sequence* of these pulses represents the intelligence to be transmitted.

Since the modulating signal only has two amplitude values, the modulated wave has only two basic frequencies. One is the unmodulated carrier frequency, and the other is the frequency that corresponds to the presence of a modulating pulse. You can see, then, that a frequency shift signal consists of an FM wave that shifts abruptly between two different frequencies.

Side-band frequencies are produced when a pulse frequency modulates a carrier, just as they are for voice or tone modulation. However, the side-band frequencies produced by a pulse are not symmetrical above and below the carrier as they are for sine-wave modulation. This is because the modulating pulses themselves are not symmetrical above and below their zero reference level. Also, a square wave, as you will learn later, is made up of a large number of odd and even harmonics of the basic square-wave frequency; so that the FM carrier is modulated by a number of higher frequencies in addition to the basic frequency. Therefore, a square wave produces many more side bands than a sine wave. Nevertheless, the number of significant (in amplitude) side-band frequencies still determines the bandwidth of the modulated wave.

FM with a triangular pulse

Another pulse shape that is often used to frequency modulate a carrier is the *triangular pulse*. Typical uses of such pulses are in radio facsimile transmission, and aircraft altimeters. Whereas a rectangular pulse increases *abruptly* to its maximum amplitude and then stays at this level until dropping sharply to zero at the end of the pulse, a triangular pulse increases *gradually*, in a *linear* way, to its maximum amplitude, and then decreases to zero in the same way. This means that when a triangular pulse is used for FM, the carrier frequency increases linearly to its highest value, and then abruptly stops increasing, and decreases linearly back to the center frequency. As you will learn later, the fact that the triangular pulse causes both a linear increase and decrease in carrier frequency is the basis for many of its uses in FM.

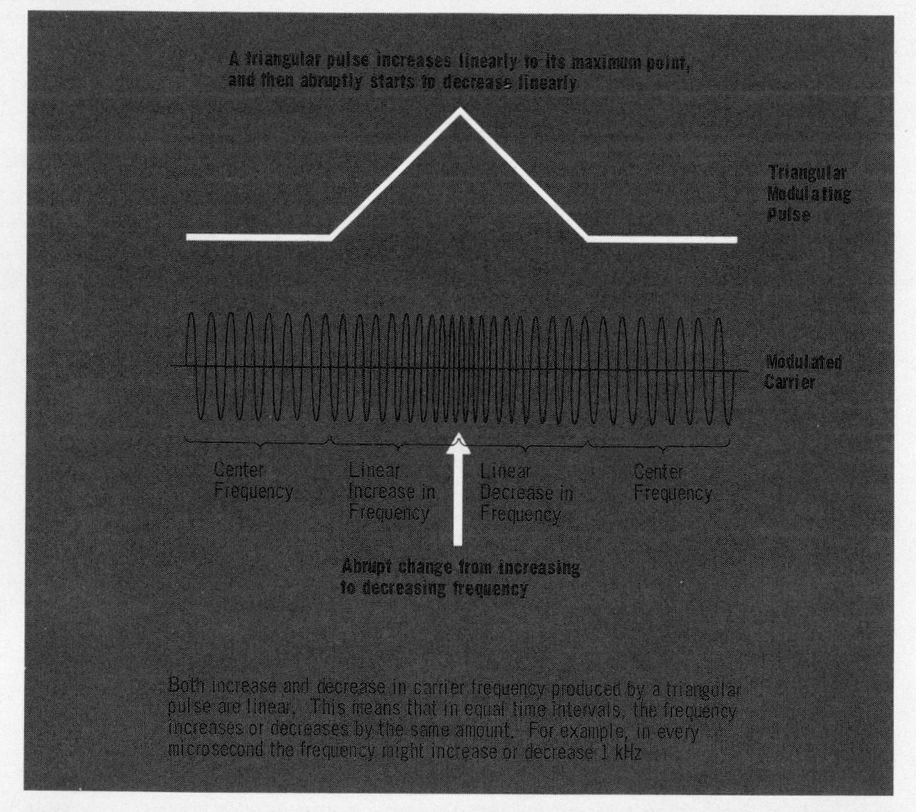

Both increase and decrease in carrier frequency produced by a triangular pulse are linear. This means that in equal time intervals, the frequency increases or decreases by the same amount. For example, in every microsecond the frequency might increase or decrease 1 kHz

The triangular pulse, like the rectangular pulse, consists of many component frequencies, called harmonics, each of which produces side-band frequencies as a result of the modulation process. These side-band frequencies, just like those produced by a rectangular pulse, are not symmetrical about the carrier center frequency, since the pulses themselves are not symmetrical about their zero reference level.

noise and FM

You will recall that a major disadvantage of AM was that noise amplitude modulates an AM signal, and, in effect, rides along with the intelligence of the signal. Noise also *amplitude modulates* FM signals. But, whereas electronic circuits cannot easily distinguish between the noise and the intelligence in an AM signal, this is relatively easy in FM signals. The *intelligence* is in the form of *frequency variations*, while the *noise* is in the form of *amplitude variations*. Therefore, by eliminating the amplitude variations from an FM signal prior to demodulation, any noise that had amplitude modulated the signal is removed.

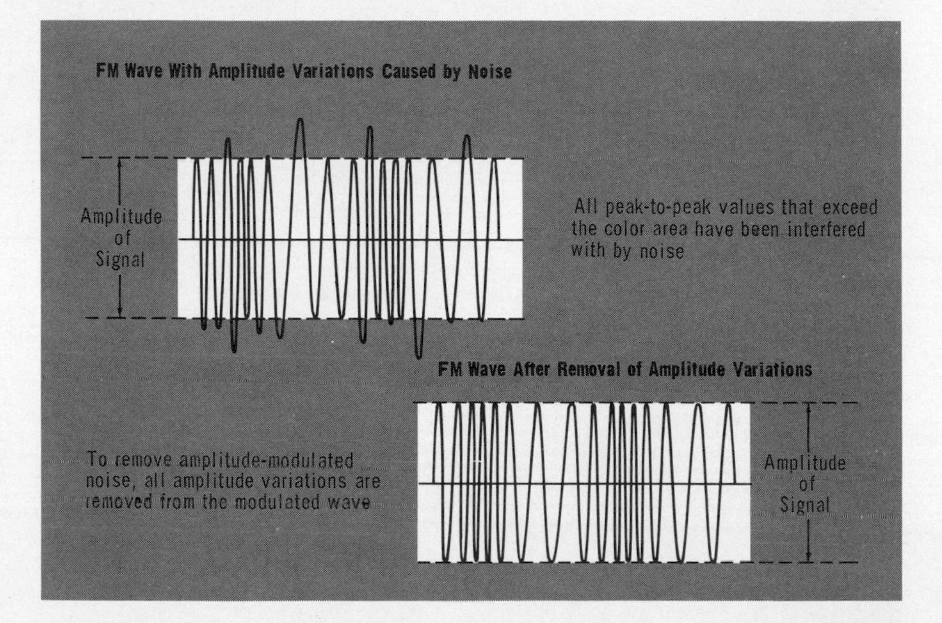

Besides producing amplitude variations, noise can also cause *frequency changes* in an FM signal. This happens when the noise pulse falls between sine waves of a carrier, and effectively widens or narrows the sine wave to make it act like a lower or higher frequency. However, the extent of the frequency changes depends not on the relative strengths of the signal and the noise, but rather on the relative values of the signal modulation index and the noise modulation index. The greater the modulation index of the signal is compared to that of the noise, the more the noise will be made ineffective, or suppressed. Thus, by using as large a *carrier deviation* as possible during modulation, the frequency changes caused by noise can be minimized. This is so, since, as you remember, the larger the carrier deviation, the greater is the value of the modulation index. With a large enough maximum carrier deviation, it is possible to almost completely suppress a noise level that is only slightly less than the signal level.

FM demodulation or detection

Demodulation, or *detection* as it is also called, is the process of recovering the original intelligence from a modulated wave. Whereas AM demodulation is accomplished with detectors that remove the carrier, leaving only the audio variations of the envelope, FM demodulation uses the modulated carrier to *reproduce* the audio signal. The circuits used for FM demodulation are sensitive to the *frequency variations* of the modulated wave, and generate a voltage that corresponds to these variations. This generated voltage thus reproduces the original modulating signal, and therefore the intelligence.

Received FM Wave

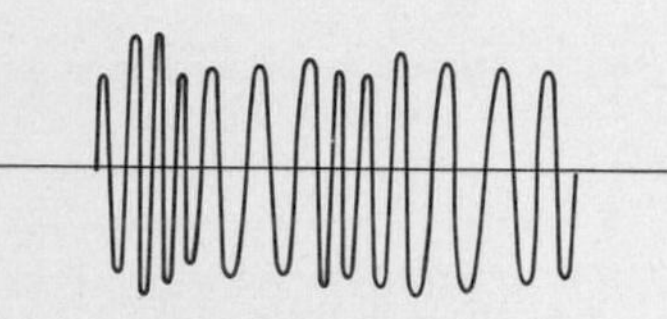

The received FM signal contains noise in the form of amplitude variations

After Noise Limiting

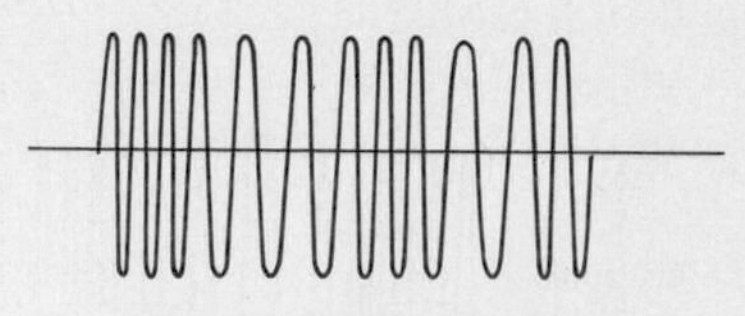

Before demodulation, the noise amplitude variations are removed from the FM wave

After Demodulation

The demodulation process then produces a signal which corresponds to the frequency variations of the modulated wave

After Deemphasis

The high frequencies of the demodulated signal are then reduced so that the resulting signal is the same as the original modulating signal

Two frequently used processes, which although are not actually part of FM demodulation, but are closely related to it, are *noise limiting* and *deemphasis.* Noise limiting takes place before demodulation, and involves the removal of amplitude variations from the modulated wave; this eliminates the noise. Deemphasis takes place after demodulation, and involves the reduction in amplitude of the high-frequency components of the signal that had been emphasized prior to modulation.

Noise limiting is the process that gives FM its big advantage over AM. Since noise occurs as an *amplitude* variation of the carrier, noise in AM is difficult to reduce because AM uses amplitude variations for its modulation. But, since FM does not, its peaks can be leveled off to remove the noise pulses.

summary

□ Preemphasis is a process whereby, before frequency modulation takes place, the higher modulating frequencies are amplified more than the lower ones. □ At the receiver, preemphasis has the result of increasing the signal-to-noise ratio of the higher-frequency components of the signal. □ Deemphasis is the opposite of preemphasis. It takes place in the receiver, and reduces the frequencies back to their normal level. □ In effect, deemphasis reduces the noise accompanying the higher frequencies.

□ Rectangular d-c pulses can be used to frequency modulate a carrier wave and produce a frequency shift signal. Such a signal shifts abruptly between two different frequencies, with the sequence of these shifts representing the intelligence being transmitted. □ Triangular pulses can also be used to frequency modulate a carrier wave. An FM signal of this type increases linearly to its highest frequency, and then abruptly stops increasing and decreases linearly back to the center frequency.

□ Noise that amplitude modulates an FM signal can easily be distinguished from the frequency variations that carry the intelligence. To eliminate the noise, all that is required is that the amplitude variations be removed. This is called noise limiting. □ Noise can also cause frequency changes in an FM signal. These can be minimized by using as large a carrier deviation as possible during modulation. □ Circuits used for the demodulation of FM signals produce a voltage that corresponds to the frequency variations of the signal.

review questions

1. What frequencies are affected by *preemphasis*?
2. Why is preemphasis used?
3. What effect does deemphasis have on the signal-to-noise ratio of an FM signal?
4. Draw waveforms that show how a frequency shift signal is produced.
5. When a rectangular pulse frequency modulates a carrier, are the side-band frequencies symmetrical about the carrier?
6. How can noise affect the frequency of an FM signal?
7. Why is an FM signal less susceptible to noise interference than an AM signal?
8. How does FM demodulation differ from AM demodulation?
9. Does deemphasis take place before or after demodulation?
10. Does preemphasis take place before or after modulation?

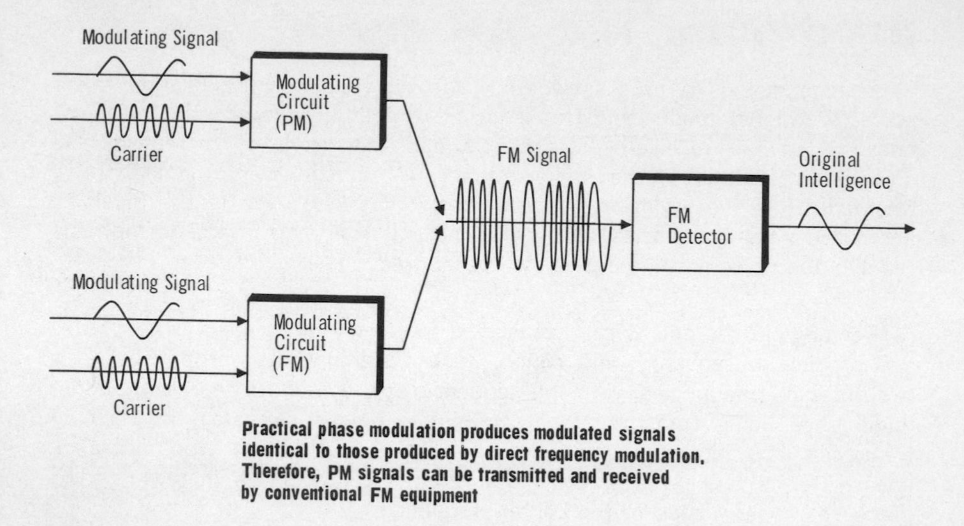

Practical phase modulation produces modulated signals identical to those produced by direct frequency modulation. Therefore, PM signals can be transmitted and received by conventional FM equipment

phase modulation

In FM, it is very important that the center frequency of the carrier stays the *same* throughout the entire demodulation process, because it is the amount of *deviation from the center frequency* that represents the intelligence. These frequency variations are detected to reproduce the original intelligence. If the carrier center frequency were to change, or *drift*, during modulation, the frequency variations would not have a *common reference* point, and the demodulated signal would be distorted.

In many of the electronic circuits used to produce FM signals, the carrier center frequency has a tendency to drift. To overcome this, a form of modulation called *phase modulation* (PM) is often used. You should understand here that phase modulation is a method of modulating a continuous-wave carrier. The *result* of phase modulation is an FM signal that is transmitted, received, and demodulated the same as FM signals you have already studied. In other words, phase modulation is an *indirect* way of producing an FM signal having a high, stable center frequency. The principle on which phase modulation works is that any change in the *phase* of a sinusoidal wave automatically causes a change in the *frequency* of the wave.

In phase modulation, the instantaneous phase of the carrier is varied from its phase at rest by an amount proportional to the amplitude of the modulating signal. The maximum phase deviation, like the maximum frequency deviation of FM, is determined by the maximum positive and negative amplitudes of the modulating signal. As the carrier is shifted in phase by the modulating signal, it also varies in frequency. These frequency variations make up what is called *equivalent FM*, and it is these frequency variations that are eventually used to recover the signal intelligence.

producing the PM signal

A significant difference between PM and direct FM is the effect of the *modulating frequency* on carrier frequency. In FM, the carrier frequency deviation depends only on the amplitude of the modulating signal, not its frequency. Thus, equal-amplitude 1-kHz and 10-kHz modulating signals will produce identical shifts in carrier frequency.

In PM, on the other hand, carrier frequency deviation is affected by *both* the *amplitude* and *frequency* of the modulating signal. Higher frequencies produce proportionately greater deviations. This means that the shift caused by the 10-kHz modulation will be ten times greater than that caused by the 1-kHz modulation.

To eliminate this greater carrier shift at higher frequencies, and make the equivalent FM produced by PM the same as directly produced by FM, the modulating signal is passed through a *correction network* prior to modulation. The correction network reduces the amplitudes of the components of the modulating signal by an amount *proportional* to their *frequency*. Thus, the higher the frequency of the component, the more its amplitude is reduced. Of course, this distorts the modulating signal. But it does so in the same, but opposite, way so that the carrier frequency deviation during modulation will be proportional to the modulating frequencies. The overall effect, then, is that all equal-amplitude components of the modulating signal, regardless of their frequency, will cause shifts in carrier frequency.

FM signals produced indirectly by PM have relatively low carrier deviations. As a result, *frequency multipliers* are used to increase the deviation to the level required for a satisfactory modulation index and bandwidth. For example, a 100-to-1 frequency multiplier will raise the maximum deviation of a signal from 150 to 15,000 kHz. All frequency components of the signal will similarly be increased in frequency by a factor of 100.

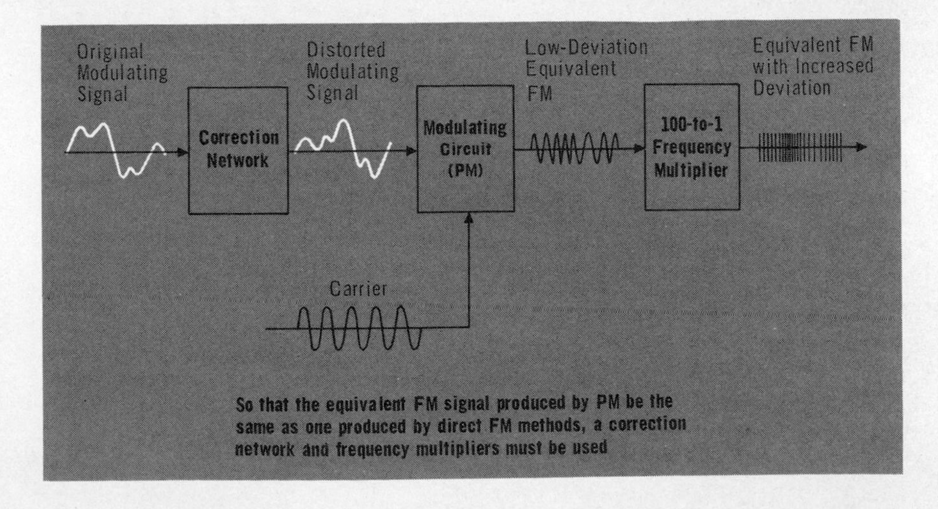

So that the equivalent FM signal produced by PM be the same as one produced by direct FM methods, a correction network and frequency multipliers must be used

signal-to-noise ratio

Since all electronic signals are actually voltages or currents that because of some varying property carry intelligence, any other voltages or currents that interact with signals will tend to *mask* the signal intelligence. These unwanted voltages or currents are grouped together under the terms *noise* and *interference*, and are produced in many ways. The most common sources of electronic noise are atmospheric static; interference generated by electrical equipment, such as motors and automobile ignition systems; and the individual components of electronic circuits. In the processing of electronic signals, some noise is always present; it can never be entirely eliminated. However, as long as the signal is sufficiently *stronger* than the noise, the signal intelligence will be unaffected.

The ratio of the signal voltage to the noise voltage is called the *signal-to-noise* ratio, S/N.

$$\frac{S}{N} = \frac{\text{signal voltage}}{\text{noise voltage}}$$

Sometimes, the signal-to-noise ratio of a signal is expressed in terms of *power* instead of voltage. In these cases, the S/N is the ratio of the signal power to the noise power. In addition, depending on the type of signal and noise under consideration, the S/N may be in terms of *peak values* of the signal and noise, or in terms of their *effective values*.

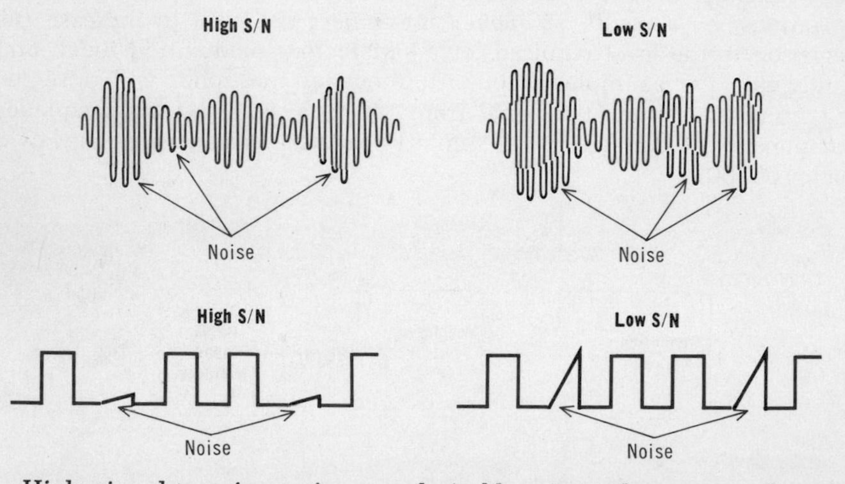

High signal-to-noise ratios are desirable, since they mean that the noise, being much weaker than the signal, will not interfere to any great extent with the signal intelligence. Low signal-to-noise ratios, on the other hand, indicate that both the signal and the noise are relatively close in value, and the intelligibility of the signal, therefore, will be partially or completely destroyed. The concept of signal-to-noise ratio applies to all electronic signals, whether they are FM, AM, dc, or ac.

pulse modulation techniques

In the discussion of the various types of modulation, you have seen how electronic signals can be made up of a series of pulses. In most of these pulse-type signals, the intelligence was carried by having the sequence of pulses correspond to some code, such as the Morse code or teletypewriter code. As a result, only intelligence that can be converted into some simple code can be carried. These pulse methods cannot be used for *complex* or *continuous* intelligence, such as voice signals. There are various other methods of pulse modulation, however, that can be used for carrying practically any type of intelligence. In all of these methods, the pulses make up the carrier, and some characteristic of the pulses is varied in accordance with the modulating signal.

Pulse modulation is an efficient means of transmitting electronic signals, since, as you learned, the power requirements of a pulse modulated signal are considerably less than those of a comparable AM or FM signal. The equipment required for pulse modulation and demodulation, however, is more complex and expensive than conventional AM and FM equipment. So pulse modulation is usually only used where its advantages outweigh the increases in size, weight, and cost of equipment.

Pulse modulation methods get their names from the way in which the pulses are varied to carry the intelligence. The most commonly used types of pulse modulation are: *pulse amplitude* modulation, *pulse width* modulation, *pulse position* modulation, and *pulse code* modulation. Sometimes, either or both pulse width and pulse position modulation are referred to as *pulse time* modulation.

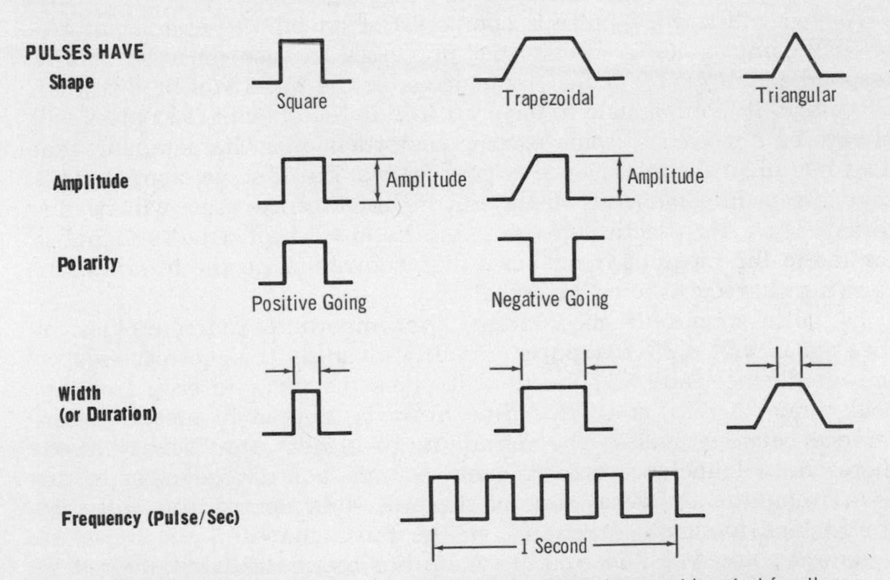

Other characteristics, such as rise and decay time, and period, are most important from the standpoint of how they affect the electronic circuits that process the pulses. They have little to do with the way in which the pulses carry intelligence

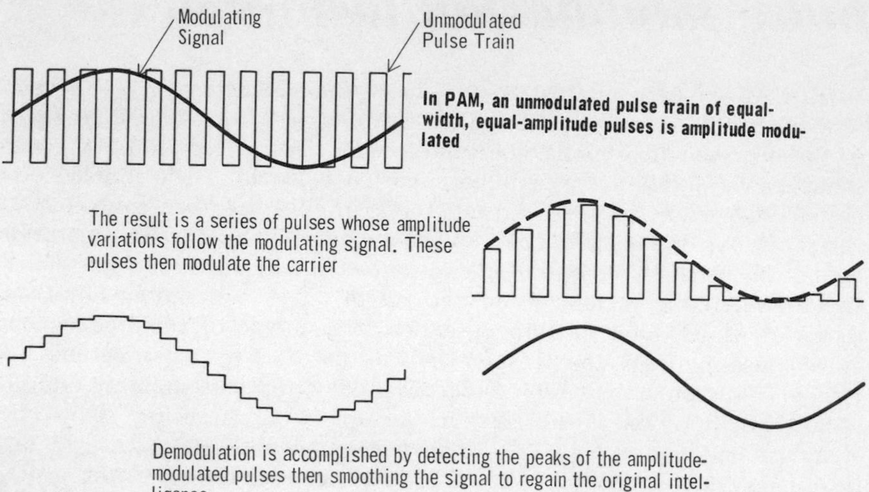

In PAM, an unmodulated pulse train of equal-width, equal-amplitude pulses is amplitude modulated

The result is a series of pulses whose amplitude variations follow the modulating signal. These pulses then modulate the carrier

Demodulation is accomplished by detecting the peaks of the amplitude-modulated pulses then smoothing the signal to regain the original intelligence

pulse amplitude modulation

As explained in the discussion of amplitude modulation, in pulse amplitude modulation (PAM), the amplitudes of the individual pulses in a *pulse train* are determined by the amplitude of the modulating signal. This breaks the modulating signal into a series of pulses whose *peaks* follow the outline of the modulating signal.

As you will learn, a pulse is composed of an *infinite* number of frequency components. A circuit that processes a pulse must be able to respond equally to *all* of the frequencies, or the pulse will be distorted. Of course, it is impossible to have an *infinite bandwidth;* so pulses will always be distorted to some extent. Distortion of a characteristic that does not affect the intelligence is permissible. But if some characteristic that affects intelligence is distorted, the intelligence itself will be distorted. Thus, for practical reasons, the bandwidth of a pulse signal is limited to the range of frequencies that contributes to the information-carrying characteristic of the signal.

In pulse amplitude modulation, the important characteristics of the signal are: (1) the pulse amplitude, since this corresponds to the intelligence; and (2) the time it takes the pulse to drop from its peak value to zero, since this time must be limited to prevent interference between pulses. The minimum bandwidth that can be used, therefore, will include those frequencies that contributed most to the peak amplitude and decay time of the pulses. Of course, this will vary from signal to signal, depending on the pulse shapes.

As was pointed out, once a pulse train has been modulated, the resulting pulse modulated signal can be used to amplitude *or* frequency modulate an r-f carrier.

pulse width modulation

In pulse amplitude modulation, the unmodulated pulse train consisted of equal-width, equal-amplitude pulses whose amplitudes were changed in accordance with the modulating signal. A similar unmodulated pulse train of equal-width, equal-amplitude pulses is used for *pulse width modulation* (PWM). However, in PWM, the *width* of the pulses *is changed* in accordance with the modulating signal. After modulation, all the pulses still have equal amplitude, but their width is proportional to the instantaneous value of the modulating signal. The leading edges of the modulated pulses correspond to the leading edges of the unmodulated pulses. As a result, the amplitude of the modulating signal is limited, since too high an amplitude would cause one pulse to run into the next.

To recover the intelligence from a PWM signal, the signal is put through any circuit whose output amplitude is proportional to the width, or time duration, of the input pulse. Wide pulses will thus produce large-amplitude outputs, and narrow pulses small-amplitude outputs. By smoothing, or averaging, the variations in these output amplitudes, the overall output of the circuit will have the same variations as the modulating signal.

In electronic literature, pulse width modulation is also often called pulse *duration* modulation (PDM) and pulse *length* modulation (PLM). Sometimes, it is also called pulse *time* modulation (PTM), but this is also used for pulse position modulation, discussed next.

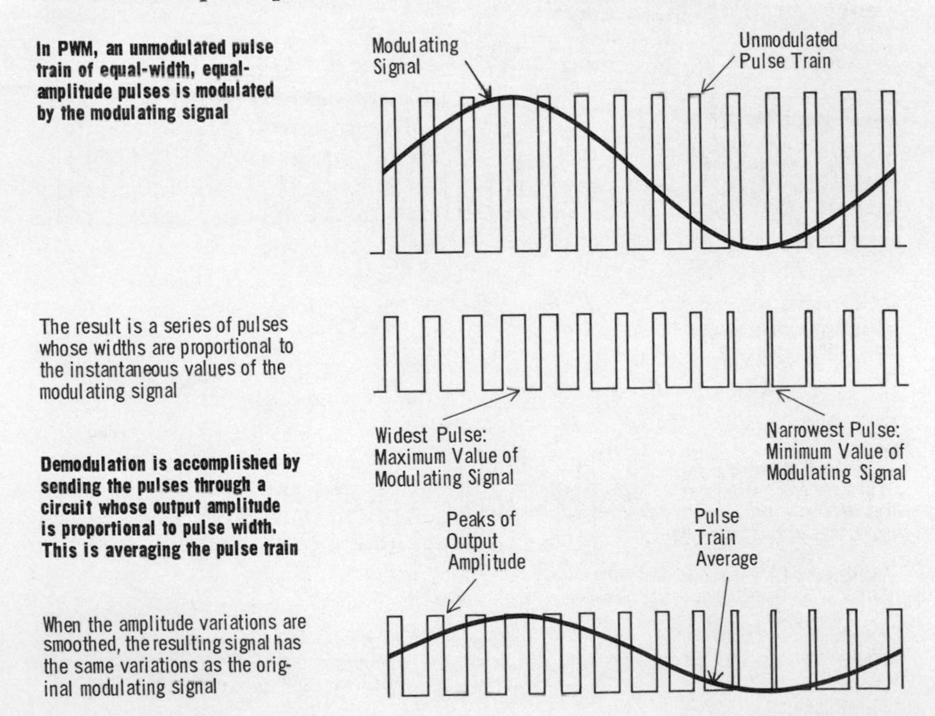

In PWM, an unmodulated pulse train of equal-width, equal-amplitude pulses is modulated by the modulating signal

The result is a series of pulses whose widths are proportional to the instantaneous values of the modulating signal

Demodulation is accomplished by sending the pulses through a circuit whose output amplitude is proportional to pulse width. This is averaging the pulse train

When the amplitude variations are smoothed, the resulting signal has the same variations as the original modulating signal

pulse position modulation

In PAM and PWM, the pulse characteristics of amplitude and width were used to add intelligence to a pulse train. Another pulse characteristic that can be used is the *position* of each pulse in the train relative to the other pulses. This is *pulse position modulation* (PPM). One way to do this is to use a circuit that *without modulation* generates a train of *equally spaced pulses*. With modulation, though, the time of occurrence of each pulse depends on the value of the modulating signal. Thus, when the modulating signal is *zero*, the pulses are generated at the *same* time as they would be without modulation. But when the modulating signal has a *positive* value, the pulses are generated *sooner*, and when it has a *negative* value, they are generated *later*. The result of the modulating process, then, is a train of *unequally spaced* pulses, with the time between the occurrence of a pulse and the point at which it would occur without modulation being proportional to the value of the modulating signal.

Because of the difficulties involved in detecting the time between the actual occurrence of a pulse and the time that it would have occurred without modulation, it is common practice in PPM to transmit equally spaced *reference pulses* with the modulated pulses. The intelligence-carrying pulses, therefore, vary in distance from the reference pulses according to the values of the modulating signal. In the demodulation process, the distance, which represents time, between each intelligence-carrying pulse and its corresponding reference pulse is detected and converted to a signal voltage.

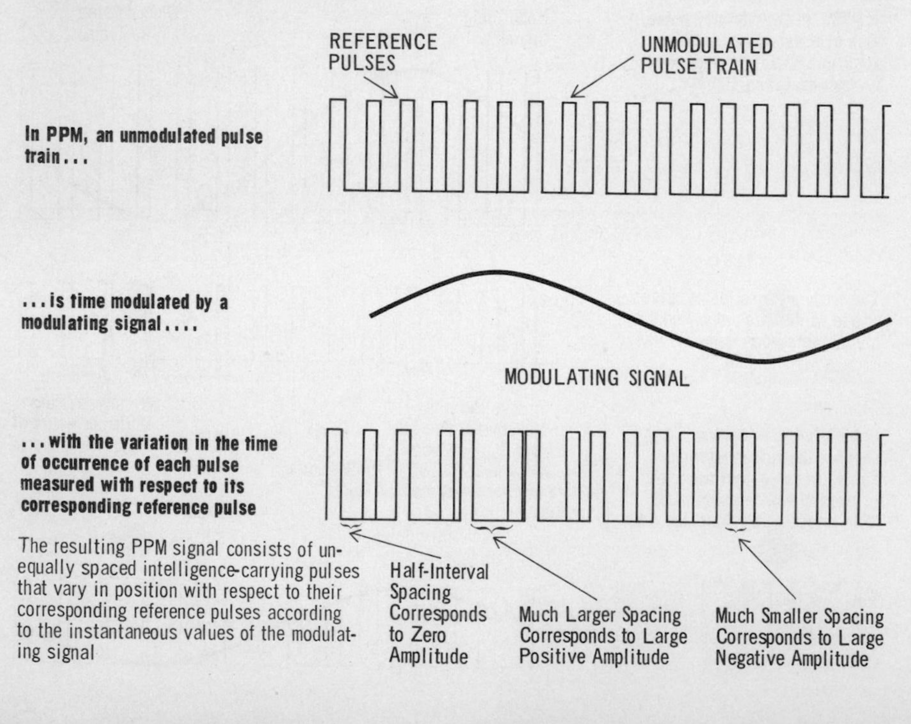

The resulting PPM signal consists of unequally spaced intelligence-carrying pulses that vary in position with respect to their corresponding reference pulses according to the instantaneous values of the modulating signal

pulse code modulation

In all of the pulse modulation techniques described, some *characteristic* was varied according to the modulation. Once the pulses have been modulated, it is important that the characteristic that represents the intelligence does not change during transmission and processing of the signal. Any change, such as in the amplitude of a PAM pulse or the width of a PWM pulse, would cause distortion of the intelligence. There is one type of pulse modulation that is much more immune to *interference* and *distortion* than the others. It is called *pulse code modulation* (PCM).

In pulse code modulation, the modulating signal is *sampled* at discrete intervals. Each time it is sampled, a *group* of pulses is produced that corresponds to the value of the modulating signal at that instant. Every pulse in the group is *identical* to every other pulse, and it is the *number* and *positions* of the pulses that represent the value of the modulating signal. The system by which the pulses in a group represent a specific value is called a *code*. By using appropriate codes, a limited number of pulses can represent a wide range of values. For example, one code, called the binary code, might be used as follows, where there are four pulse positions in each group, and 1 represents a pulse in a particular position, and 0 no pulse.

	Position					Position			
Value in Volts	1	2	3	4	Value in Volts	1	2	3	4
0	0	0	0	0	8	1	0	0	0
1	0	0	0	1	9	1	0	0	1
2	0	0	1	0	10	1	0	1	0
3	0	0	1	1	11	1	0	1	1
4	0	1	0	0	12	1	1	0	0
5	0	1	0	1	13	1	1	0	1
6	0	1	1	0	14	1	1	1	0
7	0	1	1	1	15	1	1	1	1

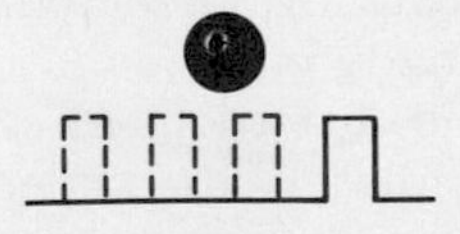

This shows how PULSE – NO PULSE groups can be used to carry digits 1 through 4

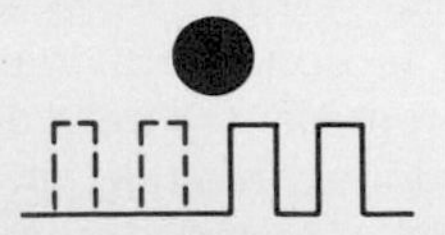

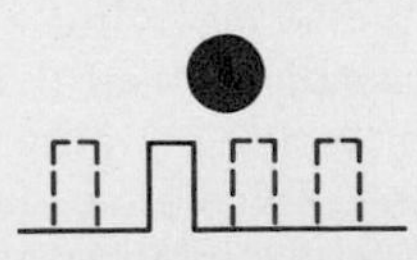

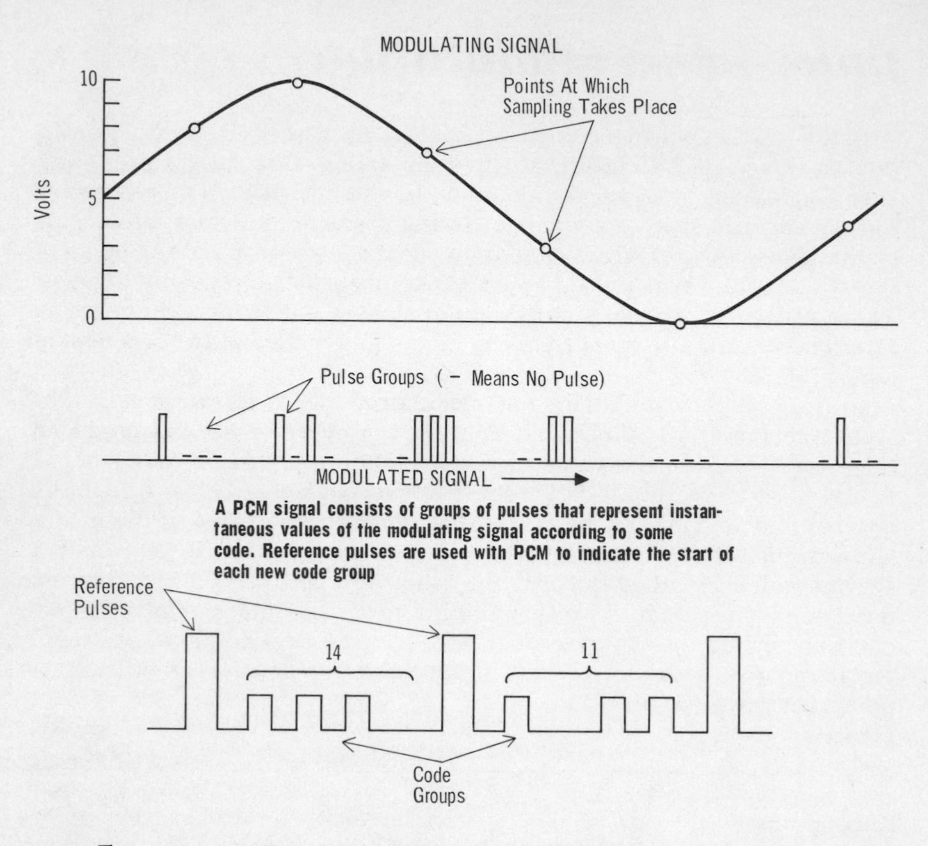

A PCM signal consists of groups of pulses that represent instantaneous values of the modulating signal according to some code. Reference pulses are used with PCM to indicate the start of each new code group

pulse code modulation (cont.)

The use of the binary code with an actual signal is illustrated. Actually, practically any satisfactory coding arrangement can be used. The important point is that the receiver uses the same code to reverse the process and recover the intelligence.

When a PCM signal is demodulated, each group of pulses goes through a circuit that, in effect, decodes the group by producing an output voltage that corresponds to the level represented by that group.

The relative immunity of PCM to interference and distortion arises from the fact that it is the *presence* or *absence* of the pulses, rather than any varying characteristic, that has to be determined to recover the intelligence. Even if the pulses are distorted, it will have no effect on their detection, unless of course the distortion is extremely severe.

Reference pulses are used with the pulse code groups to indicate to the receiving equipment that a new code group is starting. PCM is especially useful in transmitting numerical data, as is done with telemetering equipment. Also, the codes can be used to represent letters of the alphabet so that words can be transmitted, as is done with teletype. Or, as explained previously, the codes can represent relative amplitudes.

time division multiplexing

In a pulse modulated signal, some time is required between pulses. Frequently, however, the intervals are considerably longer than the pulses themselves. Such a signal is said to have a *low duty factor*. These signals can be used to advantage by having one or more pulses that represent *different* intelligence occupy the intervals. This results in a single modulated pulse train that carries *multiple* intelligence. Such a technique is called *time division multiplexing*, and makes possible the simultaneous radio transmission of more than one signal on a single r-f carrier.

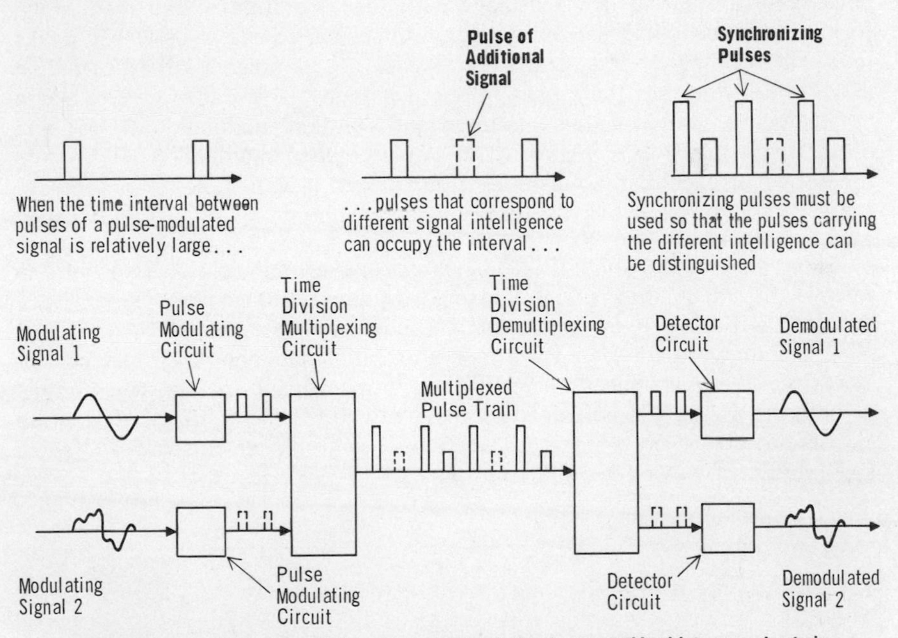

In time division multiplexing, multiple pulse-modulated signals are combined into one pulse train. The individual signals are separated after transmission by a demultiplexing circuit, and then demodulated separately

When time division multiplexing is used, the pulses of the different signals are usually distinguished by reference, or *synchronizing*, pulses. The synchronizing pulses are different from the intelligence-carrying pulses. They may have longer durations or larger amplitudes than the other pulses. The detecting circuits can then separate the pulses of the different signals, and recover the intelligence carried by each signal.

Although two separate signals are shown being multiplexed in the illustration, time division multiplexing commonly involves as many as four or more signals. In addition, synchronizing pulses are not used after every intelligence pulse. Instead, they usually separate groups of pulses that are made up of one pulse of each signal.

summary

☐ Phase modulation (PM) is an indirect way of producing an equivalent FM signal having a highly stable center frequency. ☐ In phase modulation, the instantaneous phase of the carrier is varied from its phase at rest in proportion to the amplitude of the modulating signal. This automatically causes corresponding changes in the carrier frequency. ☐ A correction network is required because the carrier deviation is affected by the frequency of the modulating signal, as well as its amplitude. ☐ Phase modulation produces relatively low carrier deviations, and so frequency multipliers are used.

☐ In pulse amplitude modulation (PAM), the amplitudes of the pulses in a pulse train are varied in accordance with the amplitude of the modulating signal. This breaks the modulating signal into a series of pulses whose peaks follow the outline of the modulating signal. ☐ In pulse width modulation (PWM), the widths of the pulses in a pulse train are varied. ☐ Pulse width modulation is also sometimes called pulse duration modulation (PDM) and pulse length modulation (PLM). ☐ In pulse position modulation (PPM), the time of occurrence of the pulses in a pulse train is varied. Reference pulses are usually required with pulse position modulation.

☐ Pulse code modulation (PCM) uses groups of pulses to represent the value of the modulating signal at sampling intervals. The number and positions of the pulses follow some specific code. ☐ Pulse code modulation has good immunity to interference and distortion, since only the presence or absence of the pulses need be determined to recover the intelligence. ☐ Time division multiplexing is a modulation technique in which two or more separate signals are placed on a single pulse train. Synchronizing pulses are used to separate the pulses of the different signals.

review questions

1. What is the relationship between PM and FM?
2. Why are frequency multipliers needed with PM?
3. Why is a *correction network* needed in PM?
4. Why are high signal-to-noise ratios desirable?
5. Draw waveforms to show how a sine-wave modulating signal can pulse-amplitude, pulse-width, and pulse-position modulate a pulse train.
6. What is the major advantage of pulse code modulation?
7. What is *time division multiplexing*?
8. Draw the waveform of a pulse train with a low duty factor.
9. Why is phase modulation used?
10. Is there any way of telling whether a received FM signal was frequency modulated or phase modulated?

complex modulation

You have learned the basic methods of modulation, in which the intelligence contained in one wave (the modulating signal) is transferred onto, or modulates, another wave (the carrier). In the descriptions of these methods, we assumed, for the most part, that the modulating signal contained a *single* intelligence; for example, a single voice signal, a single continuous tone, or maybe an interrupted tone or carrier corresponding to the Morse code. You will find, however, that many types of electronic signals are *not* this simple, because of the nature of the intelligence being carried.

In many actual signals, two or more types of *unrelated* intelligence may be transmitted on the *same* carrier, and they may or may not use the same kind of modulation. In other signals, *related* intelligence might be contained in separate modulating signals, which are used to modulate separate carriers which, again, may or may not employ the same type of modulation. These separate carriers must then be transmitted, received, and demodulated simultaneously, since the overall signal intelligence is the sum of the intelligence carried by each.

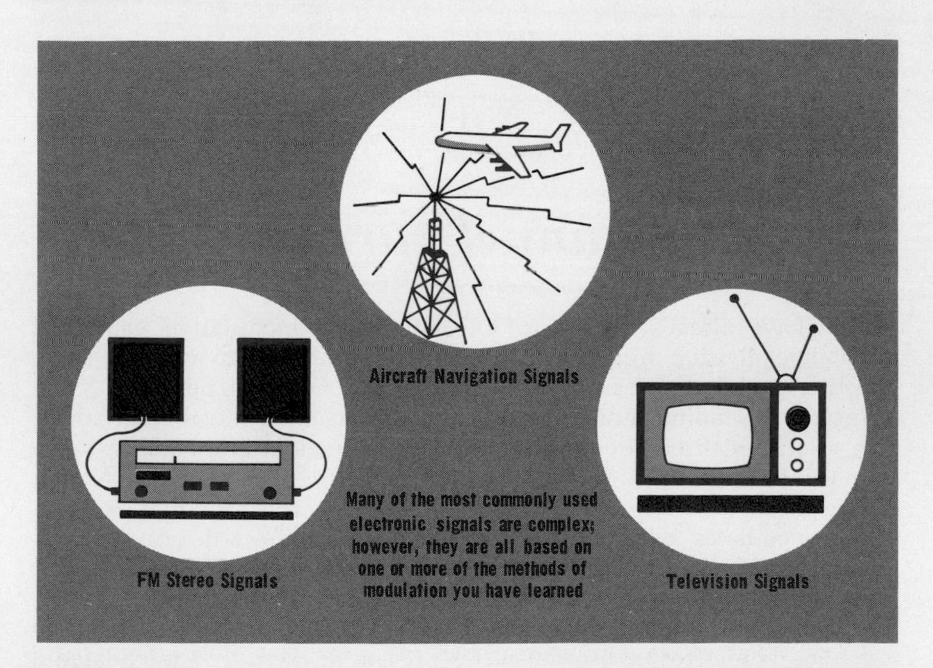

Some of the more representative types of complex modulated signals are described on the following pages. In addition to those covered, there are many more. However, once you have an understanding of how the basic modulating methods can be used to form some of the common complex signals, you should have no trouble in understanding others that you may encounter.

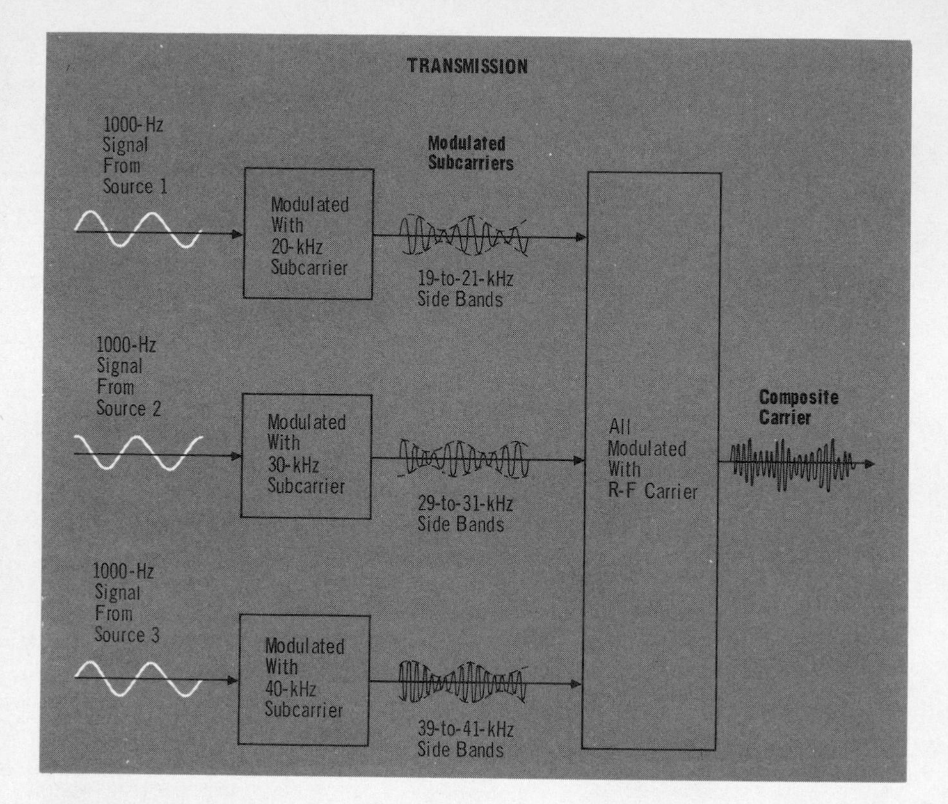

the multichannel carrier

You have already seen how by using pulse modulation combined with time division multiplexing techniques, more than one message, or other modulation, can be transmitted on a *single* carrier. This same principle of sending numerous separate messages on a common carrier can also be realized for *continuous* types of modulation, such as voice signals or tones, by a method called *frequency division multiplexing.*

Frequency division multiplexing is based on the fact that when a signal modulates a carrier, the *intelligence* is unaffected, only the *frequency* at which it is carried is changed. Thus, if two signals having the *same* frequency, or frequencies, simultaneously modulated a carrier, the two would *interact* with each other, and it would be impossible to separate them after transmission. But, if one of them first modulated a *subcarrier,* its frequency would be raised above that of the other signal. Therefore, the products of this modulation together with the other signal could then modulate an r-f carrier, and no interaction would take place between them because of their frequency difference. After transmission, they could be easily filtered out of the composite carrier, and then demodulated separately.

the multichannel carrier (cont.)

The basic principle of the multichannel carrier is, therefore, that various modulating signals each modulate a *different subcarrier*, with the subcarrier frequencies being such that the modulation products (side bands and subcarrier) of each signal occupy a different frequency range. All of these modulated subcarriers then modulate a common r-f carrier to produce a *composite* modulated carrier. Of course, this composite carrier is a very complex wave, consisting of numerous sideband frequencies. Nevertheless, by passing the composite carrier through appropriate filters after transmission, each of the modulated subcarriers can be recovered. These are then demodulated to regain the original intelligence of each of the signals.

Multichannel carriers are used to a great extent for the transmission of data from satellites and space probes. They allow an extensive amount of separate data to be transmitted in a short time from a single transmitter.

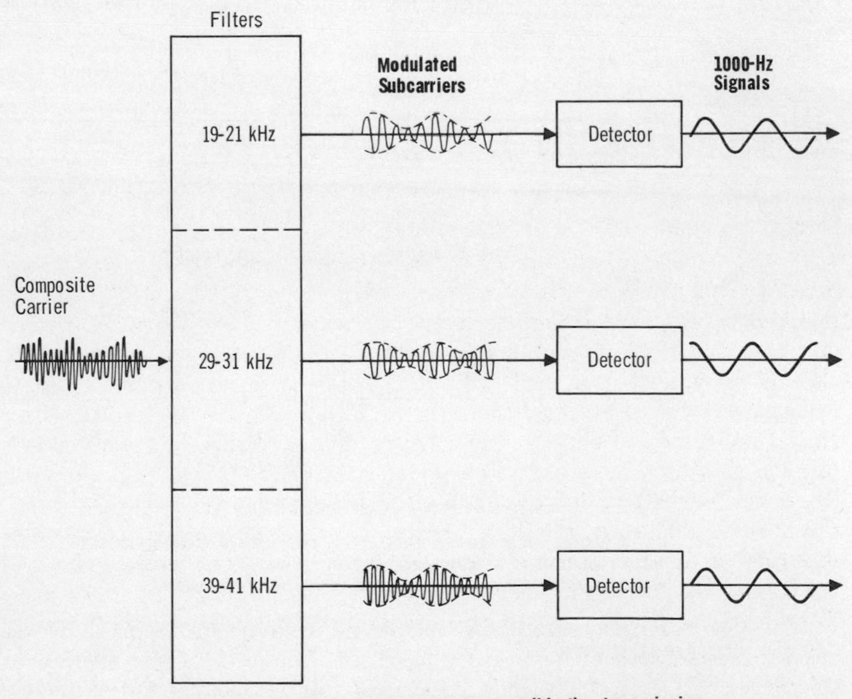

Frequency division multiplexing makes possible the transmission of numerous continuous signals on a common carrier

telephone multichannel carriers

Multichannel carriers can be used for *radio transmission* as well as for transmission by *wire* or *cable.* One of their most common uses is for the transmission of voice signals over telephone lines. By the use of multichannel carriers, many *different* voice signals can be simultaneously carried over the same transmission line without interfering with each other.

The number of channels that can be transmitted simultaneously depends on the bandwidth of the sending and receiving equipment, as well as on the bandwidths of each of the individual signals after they modulate their respective subcarrier. For example, if the equipment has a bandwidth of 200 kHz, and each modulated subcarrier has a bandwidth of 40 kHz, four channels can be sent on the main carrier. But, if the bandwidth of the modulated subcarrier is reduced, the main carrier can accommodate additional channels. This is why it is common practice to *suppress* one of the side bands of each signal after subcarrier modulation. By the removal of one side band, the bandwidth of the signal is cut in half, without interfering in any way with the intelligence being carried.

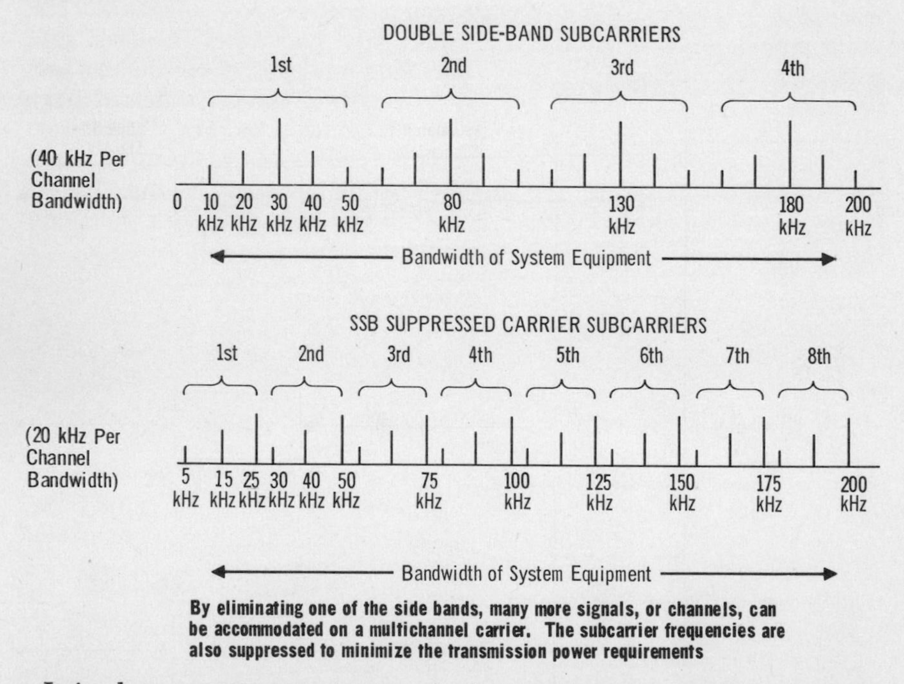

By eliminating one of the side bands, many more signals, or channels, can be accommodated on a multichannel carrier. The subcarrier frequencies are also suppressed to minimize the transmission power requirements

It is also common practice in telephone systems to suppress the subcarrier frequencies to keep the power requirements of the system at a minimum. After transmission, the SSB suppressed carrier signals are removed from the multichannel carrier by filtering, and then the subcarriers are reinserted for demodulation.

the black-and-white television signal

The signal used for the transmission of black-and-white television is a very complex signal in which the overall intelligence is made up of various individual parts or components. Basically, the transmitted television signal consists of *two separate carriers:* one modulated with the *sound,* and the other with the visual, or *video,* portion. A television receiver receives both carriers *simultaneously,* builds up, or amplifies, the level of both, and then separates the two for demodulation. The sound portion of the television signal is a standard *FM* wave.

FM is used rather than AM for the television sound because of its better immunity to noise and interference, and it is more efficient as far as transmitter power requirements are concerned. In addition, FM signals generally have a much better signal-to-noise ratio than do comparable AM signals.

Although the sound portion of a television signal is frequency modulated, the video portion is amplitude modulated onto its separate carrier. AM is used for the video portion mainly because of the possibility of *multipath reception* of the transmitted signal. This occurs when the same signal, because of the reflections from buildings, bridges, etc., reaches a receiving antenna from more than one path. Since the distance traveled by these multipath signals is usually different, different parts of the signal arrive at the antenna at the same time. For AM signals, this causes interference at the television receiver in the form of multiple images, or ghosts, on the screen. For FM signals, the interference would be much more bothersome, since it would be in the form of continuously shimmering bars on the screen.

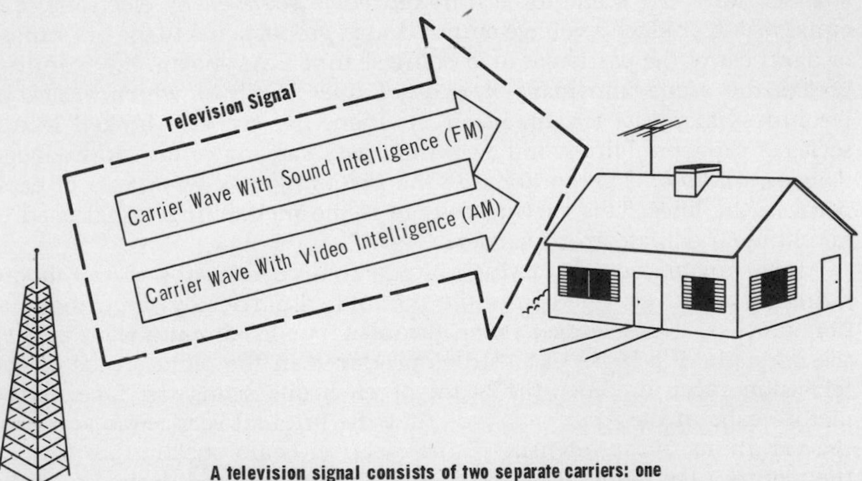

A television signal consists of two separate carriers: one frequency modulated with the sound information, and the other amplitude modulated with the video information

the video portion

The composite television carrier has the sound carrier located 4.5 megacycles above the video carrier. The sound carrier is a normal DSB transmission, but the video portion is a vestigial side-band signal. The frequencies close to the video carrier are the low video frequencies, and those farther away are the highs. Video frequencies of up to 4 MHz can be sent.

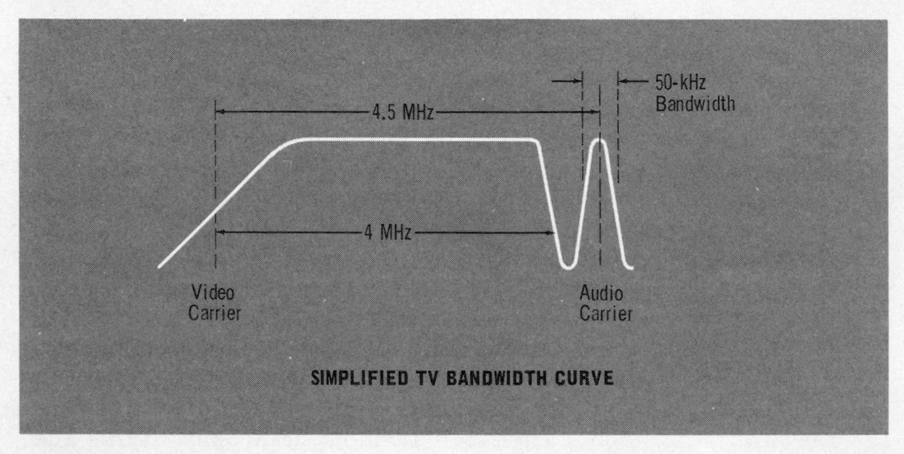

SIMPLIFIED TV BANDWIDTH CURVE

The video portion of a television signal consists of an AM carrier whose *amplitude variations* correspond to the picture or scene being transmitted. To understand how the video signal carries the visual intelligence, you must have an idea of how the signal is produced from the original scene, and then, after transmission, how the demodulated signal is converted to a reproduction of the original scene.

Essentially, the scene to be transmitted is *scanned* by electronic circuits, which produce a voltage output that is proportional to the brightness or darkness of the particular area being scanned. As shown, the scanning breaks the scene into many *horizontal lines*, each of which varies in brightness along its length. In effect, then, the scene is broken into a series of sequential lines, and a continuously varying voltage is produced whose *amplitude* is proportional to the *instantaneous brightness* of each point in the lines. This varying voltage is the modulating signal used to amplitude modulate an r-f carrier.

After transmission, a varying voltage that corresponds to the modulating signal is recovered from the modulated carrier by the demodulation process. This voltage, then, according to its amplitude, regulates the *intensity* of a beam of electrons produced in the picture tube of the television receiver. Since the beam of electrons scans the face of the picture tube in the *same sequence* that the original scene was scanned, the variations in the intensity of the electron beam striking the face of the picture tube follow the variations in brightness and darkness of the original scene.

the video portion (cont.)

Visual reproduction of the original scene occurs as a result of a *phosphorescent coating* on the face of the picture tube. When this coating is struck by the electron beam, it emits light, with the amount of light being proportional to the intensity of the beam. Thus, when the electron beam has a high intensity, which corresponds to a point of extreme brightness in the original scene, a large amount of light is emitted from the point on the picture tube struck by the beam. Similarly, points of low brightness in the original scene will result in a low-intensity electron beam striking the corresponding point on the face of the picture tube, and, hence, little light will be emitted.

The scanning of the original scene at the transmitting end and the reproduced scene at the receiving end must take place *very rapidly* so that the human eye only sees complete pictures on the television screen rather than seeing the pictures being produced line by line. In actual practice, a scene is broken into 525 individual lines, with each line being scanned in 1/15,750 of a second. In one second, therefore, the entire scene is scanned 30 times.

Effectively, then, the television video signal carries 30 complete *still pictures* each second, similar to motion picture projectors. But because of the persistency of vision of the human eye, the rapid sequence of these still pictures gives the impression of continuous action. Also, the picture is actually sent at 60 Hz; each picture is broken into two *frames*, each containing alternate lines of a picture. When they combine on the screen, the lines of one frame are interleaved with those of the other frame. This is called *interlacing*. The 60 frames then produce 30 complete pictures per second.

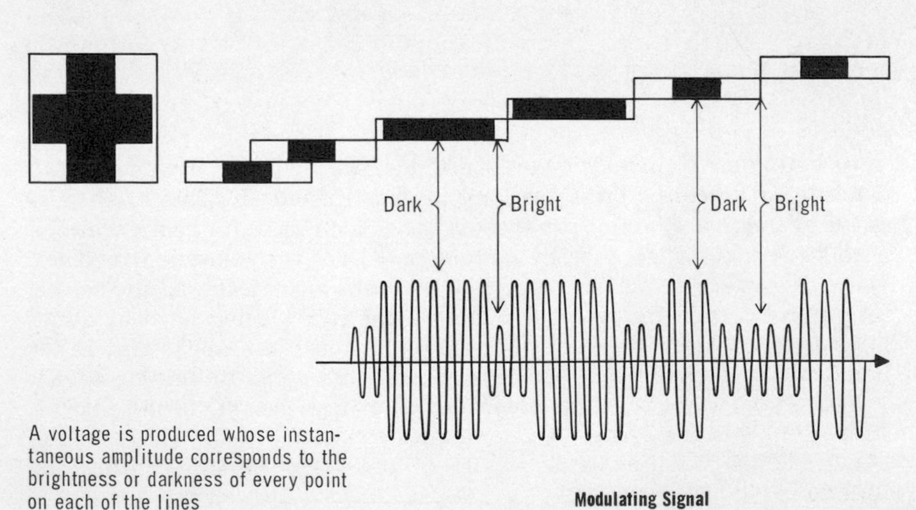

television signal fundamentals

The video signal produces the picture by means of 525 horizontal lines of varying instantaneous brightness. The scanning process that produces the picture, though, is not continuous. After each complete line there must be a time lapse before the next line begins to allow the scanning beam to return to its new starting position. In addition, after the 525th line has been scanned, there must be a time lapse to allow the beam to return to its *initial* starting point. These time lapses between lines are called *horizontal* and *vertical retrace times,* and represent periods when no picture information is being transmitted. As you will learn, it is during these times that special pulses are included in the signal to synchronize the operation of transmitting and receiving circuits.

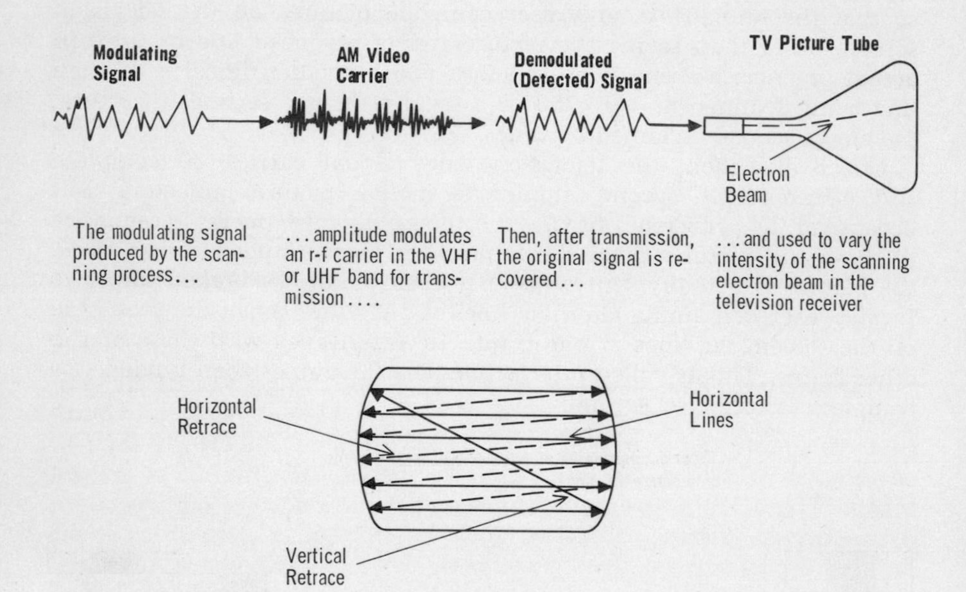

As you know from your own experience, even without a video signal, a television screen is bright as long as it is turned *on.* This brightness is called the *raster,* and is produced by the electron beam being scanned across the face of the tube by circuits inside of the television receiver. Without a video signal, the beam has a constant intensity, so the screen brightness is uniform across the face of the tube. When a video signal is received, it varies the intensity of the electron beam above and below its *no-signal* intensity, and in this way, produces a picture on the screen.

The circuits that scan the electron beam must be synchronized with those that scan the original scene at the television studio. The synchronizing pulses that are made a part of the video signal perform this function.

nonpicture portions

You have seen how the video signal carries the picture information in the form of amplitude variations that correspond to the brightness and darkness of the scene being transmitted. For proper operation of the television receiver, though, the video signal must carry more than just the picture information. It must provide a means for *cutting off* the electron beam of the receiver scanning circuits during the periods of both horizontal and vertical *retrace*. If this is not done, unwanted white lines would be produced on the TV screen during every retrace. The video signal must also provide for *synchronization* between the transmitting and receiving scanning circuits. Without such synchronization, the picture at the TV receiver would roll vertically and tear horizontally.

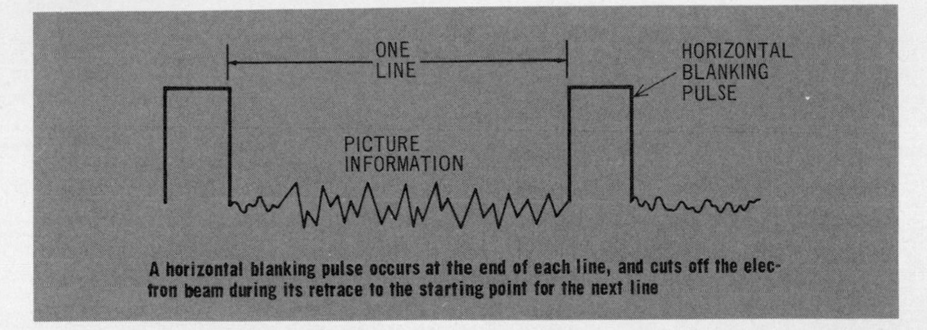

A horizontal blanking pulse occurs at the end of each line, and cuts off the electron beam during its retrace to the starting point for the next line

The electron beam of the receiver is cut off during retrace by portions of the video signal called *blanking pulses*. These pulses are rectangular in shape, and represent a signal voltage level sufficiently high to cut off the electron beam and therefore produce no brightness on the screen. There are horizontal blanking pulses for cutting off the beam during horizontal retrace, and vertical blanking pulses for cutting it off during vertical retrace. As shown, both types of blanking pulses have the same amplitude, but the *vertical* blanking pulses are wider, or of longer duration.

VERTICAL BLANKING PULSE

PICTURE INFORMATION (Last Line)

A vertical blanking pulse occurs at the end of the last horizontal line, and cuts off the electron beam during its retrace to the starting point for the first line

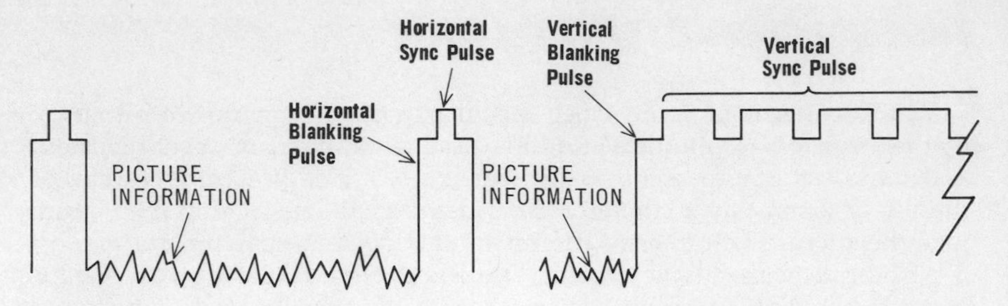

A horizontal sync pulse rides atop each horizontal blanking pulse. The sync pulse triggers the horizontal retrace, insuring that scanning of next line begins at the proper time. 18,750 blanking pulses are sent every second

A vertical sync pulse (a group of pulses) rides atop each vertical blanking pulse. The group triggers the vertical retrace, and also acts as a horizontal pulse. This group is sent 60 times per second

nonpicture portions (cont.)

Just as there are both horizontal and vertical blanking pulses, there are also horizontal and vertical *synchronizing pulses*. These are usually called *sync* pulses, and ride *on top* of the blanking pulses, as shown. The sync pulses occur during retrace, at which times no picture information is being received. This is obviously necessary, since the purpose of the sync pulses is to insure that the scanning of each line starts at the proper instant. Actually the sync pulses do this by starting, or triggering, the retrace at the end of each line. Both the horizontal and vertical sync pulses have the same amplitude, but the vertical pulses are wider. This difference in width is the basis by which the circuits in the TV receiver distinguish between the two types of sync pulses. This will be explained later.

COMPOSITE VIDEO SIGNAL

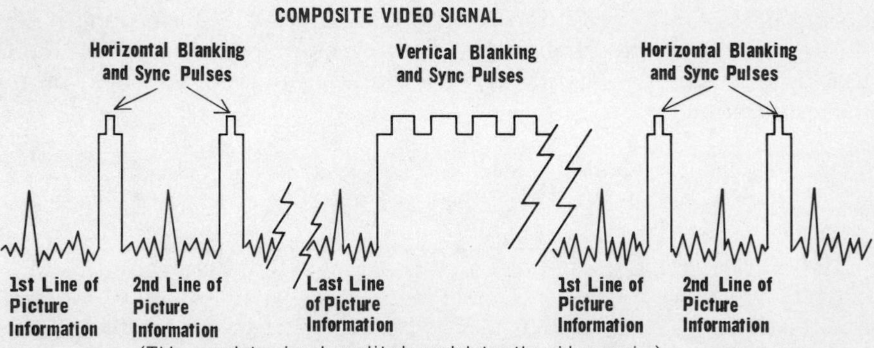

(This complete signal amplitude modulates the video carrier)

The preceding description of the video portion was of a basic nature. A complete description would involve additional elements and refinements beyond the scope of this volume

The vertical sync pulses are contained in a group of 18 special pulses that are used for both horizontal and vertical synchronization. This is explained later

bandwidth

All television signals are assigned a specific 6-MHz wide frequency channel in the VHF or UHF band by the Federal Communications Commission. There are a total of 82 channels, and they have number designations of 2 through 83. Of the 82 channels, 12 are in the VHF band (channels 2 through 13) and the remainder in the UHF band.

Within the standard channel, the FM sound carrier has a center frequency 0.25 MHz below the upper edge of the channel. The bandwidth of the sound carrier is approximately 50 kHz, which leaves about 5.7 MHz left in the band for the video carrier. If conventional double side-band modulation was used for the video signal, the highest modulating frequencies that could be transmitted would be around 2.85 MHz, which is undesirable because most of the picture detail is represented by frequencies higher than 2.85 MHz. So, instead, vestigial side-band transmission is used.

A television channel contains both the FM sound carrier and the AM video carrier in its 6-MHz bandwidth

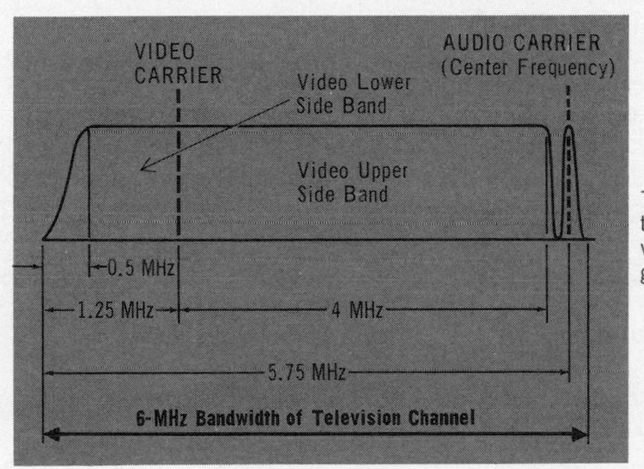

The sound carrier is a conventional FM signal, while the video carrier is of the vestigial side-band type

The bandwidths of the FM and AM portions of the overall signal are such that there is no interference or interaction between the two

The video carrier is placed 1.25 MHz above the lower edge of the channel, and all upper side-band frequencies corresponding to the maximum modulating frequencies of about 4 MHz are transmitted at their normal amplitude. The lower side-band frequencies are transmitted with varying degrees of attenuation down as far as those that are 1.25 MHz or more below the video carrier. Those frequencies below this fall outside of the assigned channel, and therefore must be *completely* eliminated before transmission.

summary

☐ Frequency division multiplexing permits multiple signals to be modulated onto a single carrier wave. Each of the signals first modulates a separate subcarrier, and these modulated subcarriers then modulate the main carrier. ☐ At the receiver, the subcarrier components of a multiplexed signal are removed from the main carrier by filters, and then demodulated separately. ☐ The number of channels that can be accommodated in a frequency multiplexed signal depends on the bandwidths of the modulating signals and the transmitting and receiving equipment.

☐ A black-and-white television signal consists of an FM sound portion and an AM video portion. The sound carrier is located 4.5 MHz above the video carrier. ☐ The video signal consists of an AM carrier whose amplitude variations correspond to the picture being sent. ☐ Scanning circuits at both the transmitting and receiving ends break the scenes being transmitted into horizontal lines. Each scene consists of 525 lines. ☐ Scanning occurs very rapidly, with each scene being scanned 30 times a second. ☐ Every picture on the television screen is broken into two frames, each containing alternate scanning lines. This is called interlacing.

☐ The brightness on a television screen when no signal is being received is called the raster. The video signal varies the intensity of the electron beam that produces the raster. ☐ The video signal also contains horizontal and vertical blanking pulses, and horizontal and vertical sync pulses. ☐ The blanking pulses cut off the scanning electron beam during horizontal and vertical retrace time. ☐ The sync pulses insure that the scanning of each line starts at the proper instant.

review questions

1. What is time division multiplexing?
2. What are *subcarriers*, and which kind of multiplexing uses them?
3. The video portion of a black-and-white television signal employs what kind of modulation?
4. How many lines are there in a television picture?
5. What is *interlacing*?
6. What is the purpose of blanking pulses?
7. What is the purpose of the sync pulses?
8. How wide is the bandwidth of a commercial television signal?
9. Is the sound carrier above or below the video carrier?
10. What are the highest video frequencies that can be transmitted by a television signal?

the color television signal

In color television, as you know, transmitted images are reproduced at the TV receiver in colors that match quite closely those of the original scene. This *color information* must therefore be transmitted by the color television signal. You are probably aware, however, that color television signals are *compatible* with both black-and-white and color television receivers. This means that a noncolor receiver produces a black-and-white picture from it, and a color receiver produces a color picture. The signal must contain color information, and also black-and-white information about the original scene. Therefore, the color television signal has components that make it quite different from the standard black-and-white signal.

The color signal is similar to the black-and-white signal in that it consists of an FM sound carrier and an AM video carrier, both contained in a 6-MHz channel. In addition, the video portion of the color signal, like that of the black-and-white signal, is made up of horizontal lines of picture information, with the individual lines followed by sync and blanking pulses. Here, though the similarity ends.

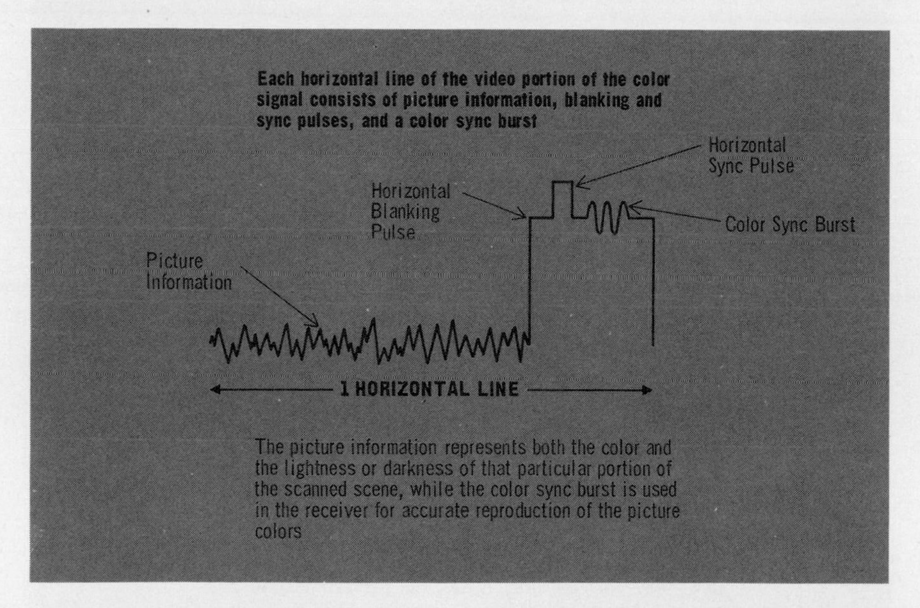

In the black-and-white signal, those portions of the signal that correspond to the picture information are merely amplitude variations that represent the brightness or darkness of the original image. But these same portions of a color signal, although still amplitude variations, are a complex representation of both the *colors* and *brightness* of the scene. Furthermore, the color signal has an additional type of sync pulse, called the *color sync burst,* which follows immediately after the horizontal sync pulses.

picture information

The picture information portion of the color video signal is a composite of color information, and brightness or darkness information. The starting point in the generation of this part of the signal is the production of *three* separate *color* signals from the image to be transmitted. Each of these three color signals is produced in the same manner as the modulating signal for black-and-white television.

One of the color signals is for *red,* and consists of a voltage whose amplitude variations follow the variations in the red content of the scene being scanned. The other two signals are for *green* and *blue,* and consist of similar voltage variations for the green and blue content of the televised scene. These three *primary colors* of red, green, and blue are the basis of the color signal, since most other colors, including white, can be obtained by mixing them in the proper ratios.

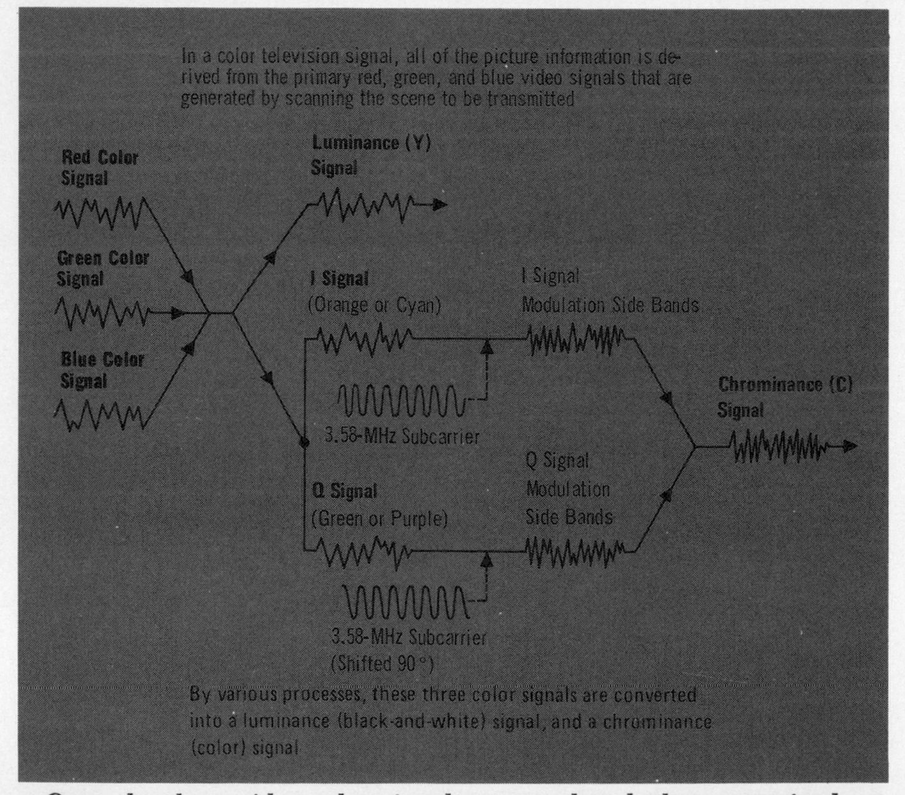

In a color television signal, all of the picture information is derived from the primary red, green, and blue video signals that are generated by scanning the scene to be transmitted

By various processes, these three color signals are converted into a luminance (black-and-white) signal, and a chrominance (color) signal

Once the three video color signals are produced, they are mixed, or combined, to produce what is called the *luminance*, or *Y*, signal. The signal corresponds to the *lightness* and *darkness* variations of the televised scene, and is similar to the modulating signal used for black-and-white television. The proportions of red, green, and blue in this "white" signal are 30 percent red, 59 percent green, and 11 percent blue.

the colorplexed video signal

The three color signals are also combined to produce two other signals, called the Q and *I* signals, besides being used to produce the luminance signal. The Q signal corresponds to the *green* or *purple* information in the picture, and the I signal corresponds to the *orange* or *cyan* information. Together, the Q and I signals contain all of the picture color information. Both the Q and I signals then amplitude modulate a 3.58-MHz carrier wave, which is called a *subcarrier*, since the results of this modulation are used to modulate the main video carrier. Actually, although it is the same 3.58-MHz subcarrier that is modulated by the Q and I signals, the subcarrier is shifted in phase 90 degrees before modulation by the Q signal. This allows both the Q and I modulation to be carried on the same subcarrier and be distinguishable from each other. The side bands produced by the Q and I modulation are added vectorially to form what is called the *chrominance, or C,* signal. The 3.58-MHz subcarrier is suppressed when the side bands are combined.

The luminance (Y) and chrominance (C) signals, which are produced from the basic red, green, and blue video color signals, contain all of the picture information to be transmitted. They are combined into one signal by being added in such a way that the variations in the *average* value of the resulting signal represent the *luminance* variations, while the *instantaneous* variations represent the *chrominance* information. The composite signal is the picture information portion of the video modulating signal, and together with the blanking, sync, and color burst sync pulses make up what is called the *colorplexed* video signal. This total video signal amplitude modulates the video carrier for transmission to the receiver.

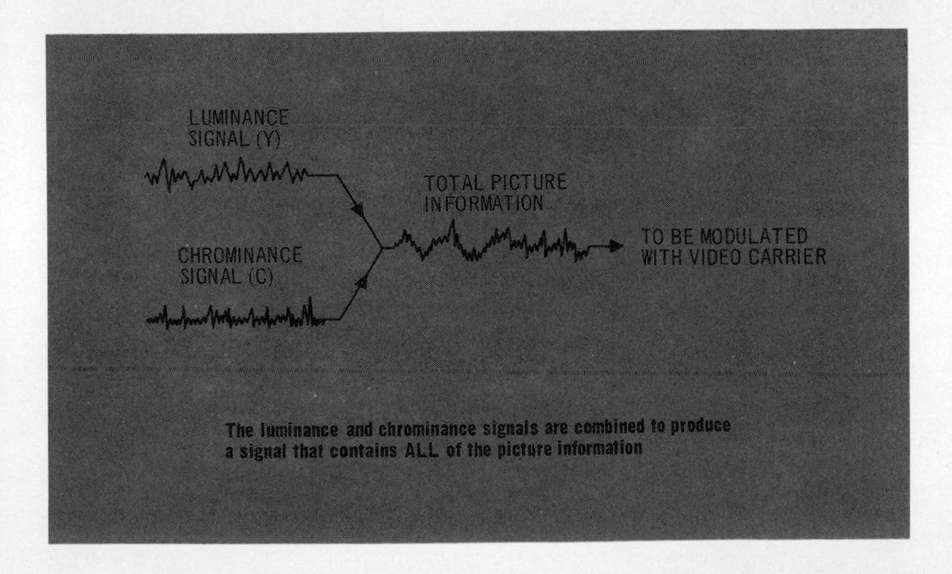

The luminance and chrominance signals are combined to produce a signal that contains ALL of the picture information

the colorplexed video signal (cont.)

Each color sync burst consists of a few cycles of the unmodulated, 3.58-MHz subcarrier used to produce the Q and I signals. You recall that the subcarrier is *suppressed* after the Q and I signals are generated. This means that the subcarrier must be reinserted in the receiver for detection of the Q and I signals. The color sync burst is used at the receiver to synchronize the phase of the reinserted subcarrier with that of the original subcarrier at the transmitter.

After the colorplexed video signal has been removed from its carrier by detection in the receiver, the way in which the signal is processed depends on whether the receiver is designed for color, or just black and white. In black-and-white receivers, only the luminance (Y) signal is detected from the picture information portion of the video signal. This is then processed, along with the blanking and sync pulses, the same as a standard black-and-white signal.

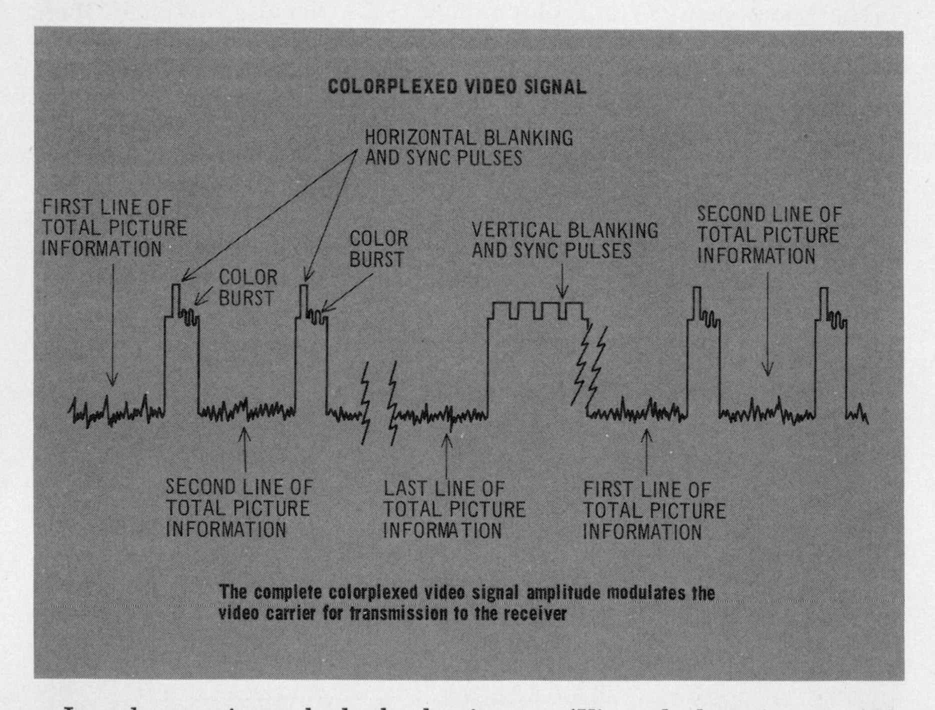

The complete colorplexed video signal amplitude modulates the video carrier for transmission to the receiver

In color receivers, both the luminance (Y) and chrominance (C) signals are detected, and together with the blanking, sync, and color burst sync pulses, are used to reproduce the color picture. The chrominance signal provides the color variations for the picture, while the luminance signal provides the variations in intensity of or brightness of the colors.

the FM stereo multiplex signal

Everyone is familiar to some extent with the stereophonic (*stereo*) reproduction of sound. Basically, it involves the use of two microphones rather than one to pick up the sound being recorded or transmitted. The two microphones are spaced some distance *apart,* and so receive somewhat different sound waves. For example, in the stereo pickup of an orchestra, one microphone might be closer to the violin section than the other microphone. Although the sound waves picked up by both would be similar, the amplitudes of the high-frequency sound components received by the microphone closest to the violin section would be larger than the same frequency components received by the other microphone, since violins are essentially high-frequency instruments.

By having the microphones produce *separate* signals and keeping the signals separate during recording or transmission, it is possible to eventually apply the two signals to separate speakers for reproducing the original sound. In this way, one speaker reproduces sound picked up by one of the microphones, and the other speaker the sound from the second microphone. The sound coming from the two speakers, then, has *depth* or *directional* qualities, somewhat similar to the three-dimensional reproduction of visual scenes.

Despite its popular acceptance, the stereo reproduction of sound was for years limited to use on phonograph and tape recordings. Radio transmission of stereo sound was impractical, since to keep the two sound signals separate, they had to be transmitted separately, as well as processed by radio receivers separately. But, finally, a system was approved by the FCC in which stereo sound can be transmitted on a single FM carrier. This type of transmission is called *FM stereo multiplex,* and is coming into relatively widespread use.

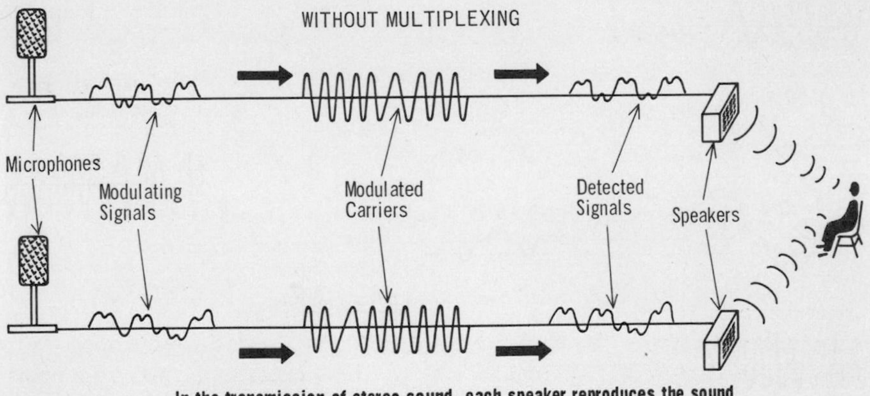

In the transmission of stereo sound, each speaker reproduces the sound picked up by one of the microphones

Without multiplexing, the transmission of stereo sound would require completely separate transmission facilities for each signal

components of the FM stereo multiplex signal

One of the principal advantages of FM stereo multiplexing is that it is compatible with FM receivers that are equipped for stereo, as well as with those that are not. Stereo receivers reproduce the transmitted signal as stereo sound. Conventional (monophonic) receivers, on the other hand, reproduce it as a conventional FM sound signal.

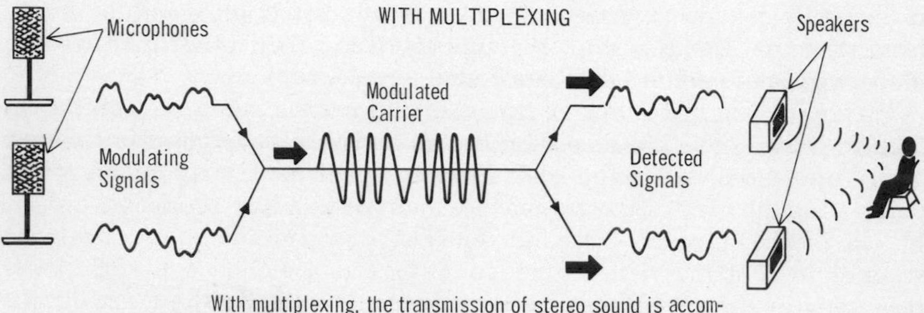

With multiplexing, the transmission of stereo sound is accomplished by placing both signals on a common carrier

The generation of the FM stereo multiplex signal begins with the audio signals produced by the two microphones. These signals are designated *L* and *R*, based on the relative positions, left (L) and right (R), of the microphones. The L and R audio signals are applied to a circuit, called a *matrix*, which develops two *new* signals.

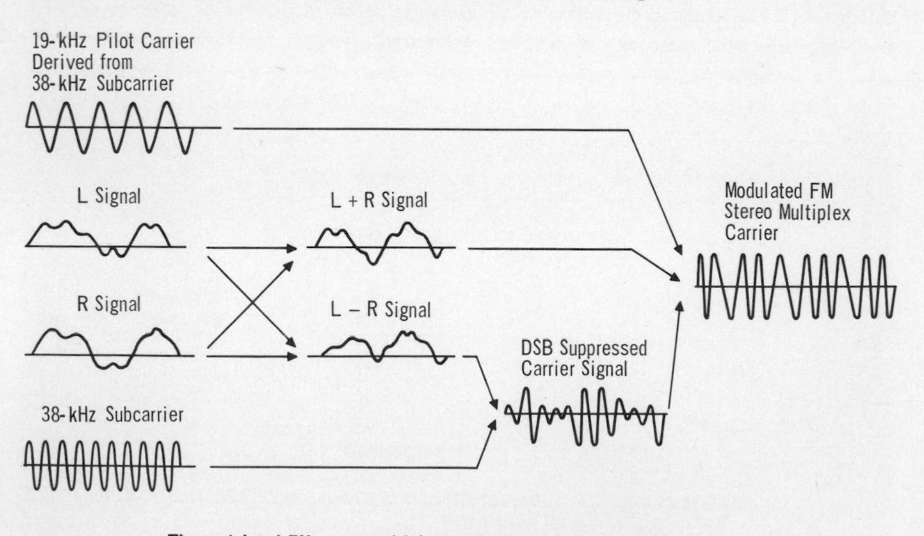

The modulated FM stereo multiplex carrier has many components. At the receiver, these components can be recovered and processed so that the original L and R signals are reproduced

components of the FM stereo multiplex signal (cont.)

One of these new signals corresponds to the instantaneous *sum* of the L and R signals, and is called the *L + R signal.* The other new signal is the *L — R signal,* since it corresponds to the instantaneous *difference* of the L and R signals. The L — R signal then *amplitude* modulates a 38-kHz subcarrier to produce side bands above and below 38 kHz. The 38-kHz subcarrier frequency is *suppressed* after the modulation takes place. Both the L + R signal, and the AM side bands produced by modulating the L – R signal with the subcarrier are then used to frequency modulate an r-f carrier for transmission.

When the original L and R signals are produced, they are limited in bandwidth to 0 to 15 kHz. All higher audio frequencies are filtered out. Therefore, the L + R signal that modulates the FM carrier has a bandwidth from 0 to 15 kHz, while the lower side band produced by the L – R modulation goes from 23 to 38 kHz, and the upper side band from 38 to 53 kHz. The intelligence of the L + R and L – R signals are thus separated by 8 kHz (15 to 23 kHz), making it easy to distinguish them after they are separated from the FM carrier following transmission.

Since the 38-kHz subcarrier was suppressed after being used for the L – R modulation, it must be reinserted at the receiver for demodulation. To make this reinsertion possible, a 19-kHz *pilot carrier* is also transmitted on the FM carrier. In the receiver, the pilot carrier is frequency doubled to 38 kHz, and then used to synchronize the 38-kHz demodulation subcarrier with the modulation subcarrier at the transmitter. The pilot carrier is sent at 19 rather than 38 kHz, since in this way it falls into the signal frequency spectrum at a point where no signal intelligence is located. This permits it to be easily detected at the receiver.

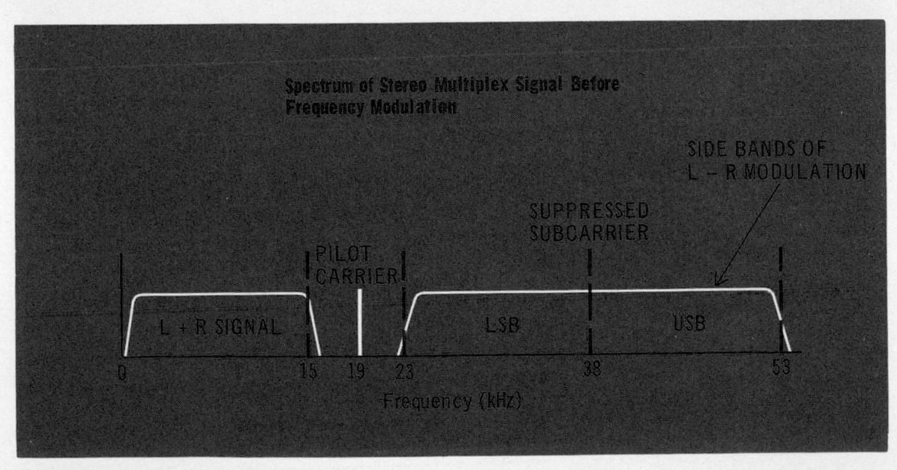

recovering the intelligence

When an FM stereo multiplex signal is received by an FM receiver, the intelligence is removed from the carrier by the demodulation process. This intelligence consists of the L + R signal, the side bands of the L − R AM signal, and the 19-kHz pilot carrier. If the receiver is not equipped for stereo, it responds *only* to the L + R signal, and processes this as a normal monophonic signal.

In a receiver that is equipped for stereo, the L − R signal is recovered by combining the L − R modulation side bands with a 38-kHz carrier, and then removing the original L − R signal. The 38-kHz carrier is generated in the receiver, and uses the 19-kHz pilot carrier for synchronization. The L − R signal and the L + R signal are then combined in a matrix circuit, similar to the one used at the transmitting end of the system. In the matrix, the L + R and L − R signals are *added* to produce the original *L signal.* Also in the matrix, the L + R and L − R signals are *subtracted* to produce the original *R signal.* These sum and difference signals are then sent to different speakers. Thus, one speaker reproduces the sound picked up by one microphone; the other speaker reproduces the sound picked up by the other microphone.

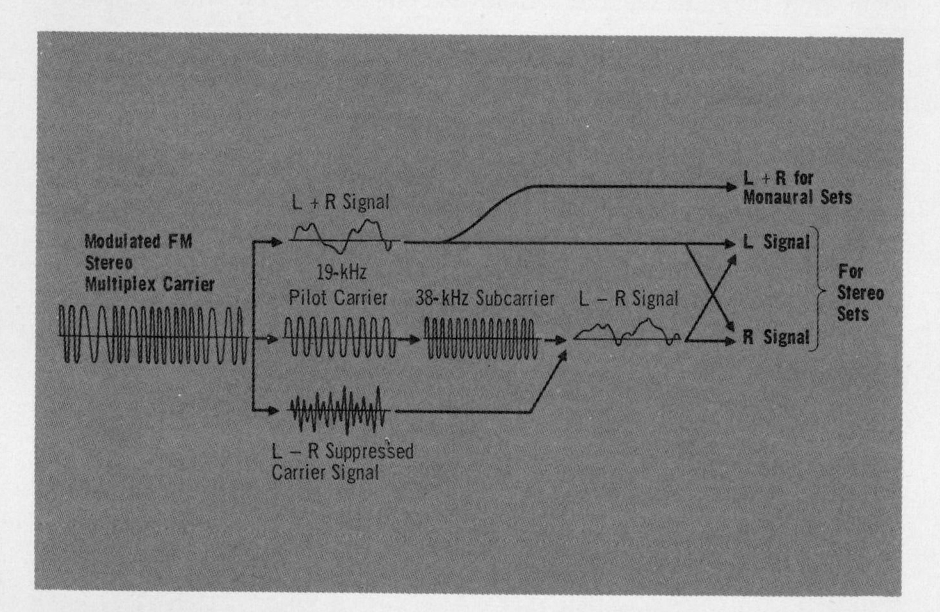

The addition and subtraction of the L + R and L − R signals is a matter of standard algebraic addition and subtraction. The sum of the two signals is

$$\begin{array}{r} L + R \\ +(L - R) \\ \hline 2L + 0 \end{array} \qquad \text{or just} \qquad 2L$$

recovering the intelligence (cont.)

When the two signals were added, the R portions cancel each other, while the L portions reinforce each other to produce an L signal with twice the amplitude. The same type of relationship holds for the difference of the L + R and L − R signals, if you remember that in algebraic subtraction, you change the signs of the terms in the subtrahend, and then add. Thus,

$$\begin{array}{r} L + R \\ -(L - R) \\ \hline \end{array} \quad \text{can be written as} \quad \begin{array}{r} L + R \\ +(-L + R) \\ \hline 0 + 2R \end{array} \quad \text{or just} \quad 2R$$

The effect of algebraic addition and subtraction of signals can be seen by analyzing these simple waveforms

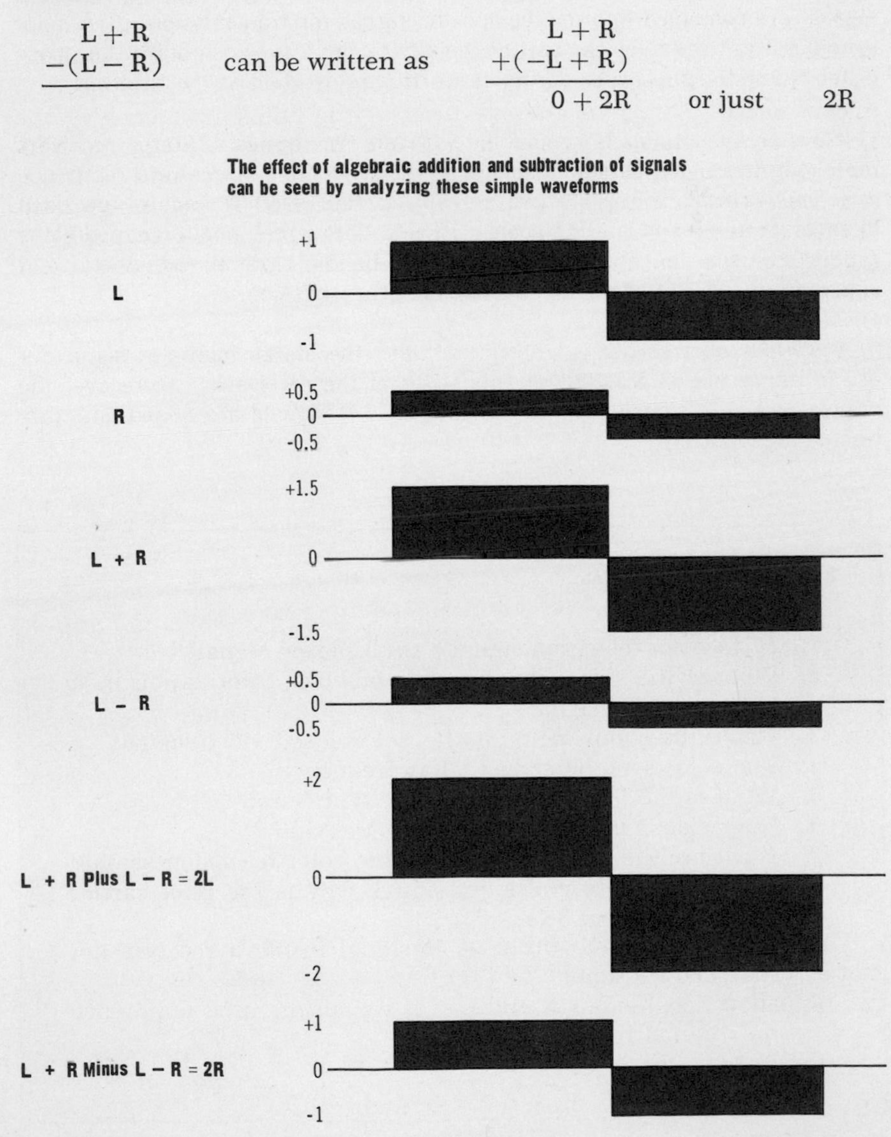

summary

□ The color television signal is compatible: a noncolor receiver produces a black-and-white picture from it. □ Red, blue, and green video signals are produced from the image to be transmitted. These three signals are combined to produce the luminance, or Y, signal, which corresponds to the lightness and darkness variations of the scene being transmitted. □ The three color signals are also combined to produce the Q and I signals. □ The I signal modulates a 3.58-MHz subcarrier, and the Q signal modulates the same subcarrier shifted 90 degrees in phase. The resulting side bands are added vectorially to produce the chrominance, or C, signal. □ The Y and C signals are combined into one composite signal for transmission. □ A color sync burst is transmitted as part of the video signal to synchronize the phase of the 3.58-MHz subcarrier during subcarrier reinsertion at the receiver.

□ FM stereo multiplex is compatible with all FM receivers. Stereo receivers reproduce the signal as stereo sound, while conventional receivers reproduce it as monophonic sound. □ Two signals, L (left) and R (right), are used to generate L + R and L − R signals. □ The L − R signal amplitude modulates a 38-kHz subcarrier, and the resulting side bands, together with the L + R signal, frequency modulate an r-f carrier for transmission.

□ A 19-kHz pilot carrier is transmitted with the stereo multiplex signal for use in reinserting of the 38-kHz subcarrier at the receiver. □ To recover the original L and R signals, the L + R and L − R signals are added and subtracted algebraically.

review questions

1. Is the color television signal a multiplexed signal?
2. What is the signal that carries the color information in a color television signal?
3. Name the components of the colorplexed video signal.
4. The color sync burst has what frequency?
5. What is the purpose of the color sync burst?
6. How wide is the color television bandwidth?
7. What are the primary colors of the color television signal?
8. In an FM stereo multiplex signal, why is the pilot carrier 19 kHz instead of 38 kHz?
9. What happens to the L − R signal inside a receiver not equipped for stereo?
10. Show how the L + R and L − R signals are used to produce the original L and R signals.

VAR uses four separate signals, each transmitted by its own antenna in a 180-degree arc

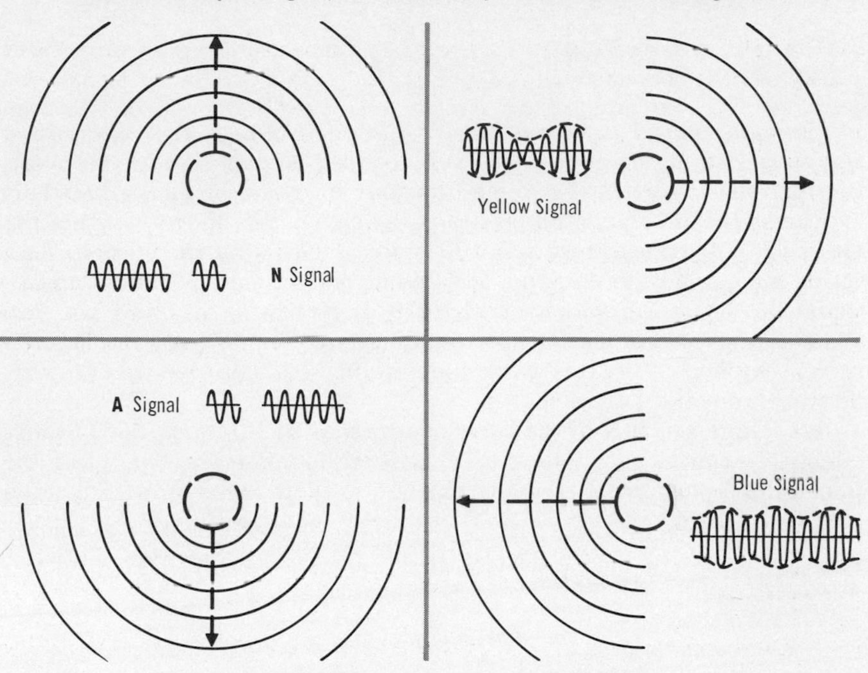

aircraft navigation signals

Ever since aircraft have been designed to fly at heights that make *visual* navigation by means of rivers, roads, and other landmarks impossible, pilots have had to rely almost exclusively on electronic signals and equipment for in-flight navigation. The electronic signals used for aircraft navigation are of many types, with their exact nature depending on the particular type of navigation for which they are used. Some of these signals, and therefore their associated electronic equipment, are quite complex; others are no more than a combination of two relatively simple signals whose relative presence or absence represents the signal intelligence.

One of the earlier aircraft navigation aids is called the *visual-aural range* (VAR). In this system, *four* separate signals are transmitted from directional antennas. Each of the four signals is on a separate carrier, but all of the carriers have the same frequency. This allows the signals to be independent, and yet be received simultaneously. Two of the signals represent the Morse code for the letters N and A, while the other two signals are 90- and 150-Hz tones. A receiver in the aircraft demodulates the signal and converts the N and A signals into *audible* tones, listened to by the pilot; while the 90- and 150-Hz tones are converted to d-c voltages, which cause an indicating instrument to indicate either blue or yellow.

aircraft navigation signals (cont.)

The area around the transmitting station in which the transmitted signals are strong enough to be used is called a *range;* and, effectively, the four signals divide the range into four quarters, or *quadrants.* This is done by the *directional* properties of the signals. As shown, each of the signals is only transmitted in a 180-degree arc around the station. As a result, each quadrant contains a *different pair* of signals. Thus, an aircraft can determine which quadrant it is in on the basis of the two signals received. For example, if a pilot heard the code for the letter A in his earphones and his instrument indicated blue, he would know he was in quadrant 4, as shown.

At the intersections of the color quadrants, the 90- and 150-Hz tones effectively cancel each other; so anywhere along these two lines the pilot's instrument reads neither blue nor yellow. This zero indication

The four signals divide the area around the transmitting station into four quadrants, with each quadrant containing a unique pair of signals

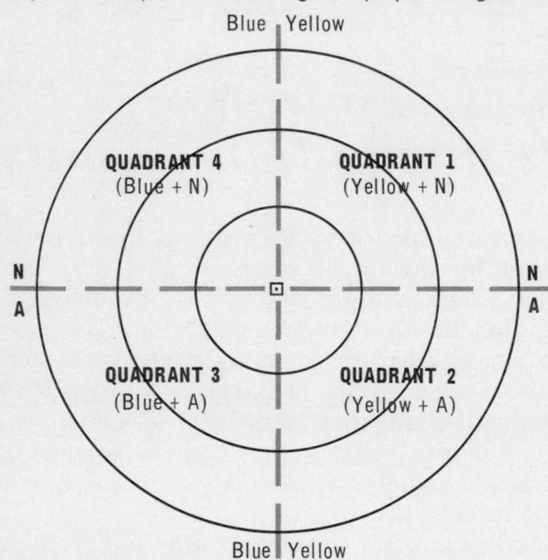

The two audible signals are heard in earphones, and the two visual signals actuate an indicating instrument whose pointer points to either a yellow or blue area

can be used for navigating *directly towards* or *directly away* from the station. Also, at the intersections, the Morse code for the letter A, which is dot–dash, and the code for letter N, which is dash–dot, combine to give a continuous tone.

Of course, with this as with most navigation systems, various charts and maps must be used to properly interpret the signals in terms of geographic directions and areas.

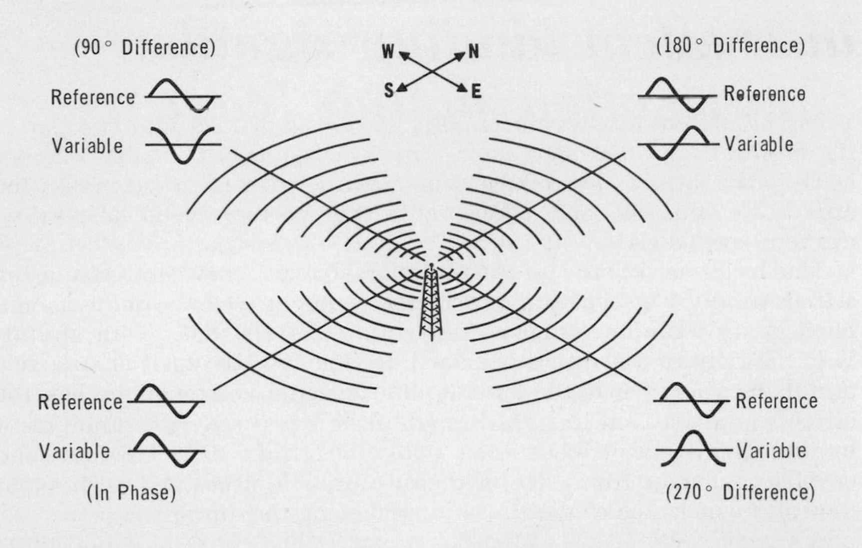

In the VOR system, the amount of phase difference between the two signals depends on the direction of transmission

By detecting and measuring the phase difference, the exact bearing of an aircraft from the station can be determined

the visual omnirange

In the VAR navigation system, both *visual* and *audible* signals tell a pilot what *quadrant* he is in. A more sophisticated system, the *visual omnirange* (VOR), permits a pilot to read his *exact bearing* on one instrument. The VOR system uses two signals, one of which has a *fixed phase*, and the other has a phase that depends on the *direction* of transmission. Thus, the first, or *reference phase,* signal has the same phase no matter where it is received in the entire 360-degree arc around the station. The *variable phase* signal, on the other hand, has a different phase for every direction of transmission. It is in phase with the reference signal in the *due south* direction, and becomes increasingly out of phase in a clockwise direction from due south.

Both the reference signal and the variable phase signal are 30-Hz tones. The 30-Hz reference signal first frequency modulates a 9960 *subcarrier*, and then this FM subcarrier amplitude modulates its r-f carrier for transmission. The 30-Hz variable phase signal amplitude modulates its carrier directly. The r-f carrier is somewhere in the frequency range of 112 to 118 MHz. Putting the reference signal on a 9960 subcarrier makes it possible to easily separate the two signals from the r-f carrier by filtering after transmission. FM demodulation of the reference signal and AM demodulation of the variable phase signal then restores the two original 30-Hz tones so they can be compared in phase. The phase difference causes the heading indicator to show the exact bearing of the plane from the airport.

instrument landing signals

Essentially, an instrument landing system is divided into two parts: the *localizer,* and the *glide slope*. The localizer tells the pilot whether he is to the *right* or *left* of the center of the runway as he makes his approach, while the glide slope tells him whether he is *descending* at the proper angle.

The localizer consists of two carriers having the same frequency and transmitted by separate *directional* antennas. One carrier is modulated by a 90-Hz tone, and the other by a 150-Hz tone. If an aircraft is to the right of the runway center line, the 150-Hz signal is stronger; and if the aircraft is to the left, the 90-Hz signal is stronger. When the aircraft is directly on line, the signals have equal strength. Equipment in the aircraft demodulates the two tones from their carriers, and develops a voltage from each of them to drive a *visual* instrument that indicates which side of the runway center line the aircraft is on.

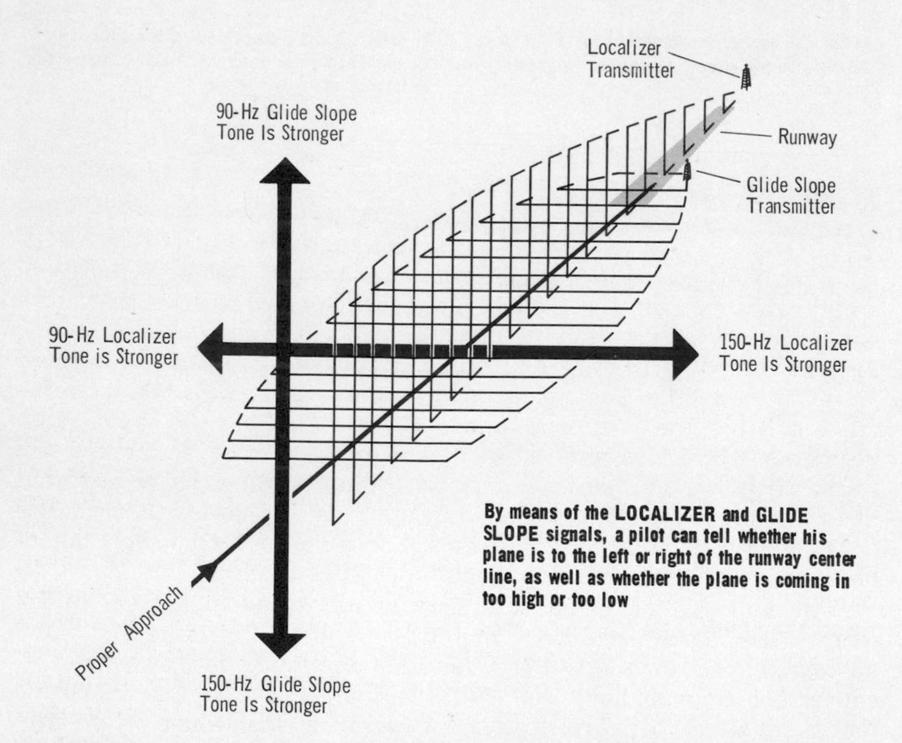

By means of the LOCALIZER and GLIDE SLOPE signals, a pilot can tell whether his plane is to the left or right of the runway center line, as well as whether the plane is coming in too high or too low

The glide slope is similar to the localizer, except that it produces a somewhat horizontal radiation pattern, as shown. When an aircraft is above the proper glide path, the 90-Hz signal is stronger; and when it is below, the 150-Hz signal predominates. These signals are demodulated in the aircraft and sent to another pointer in the same instrument used for the localizer.

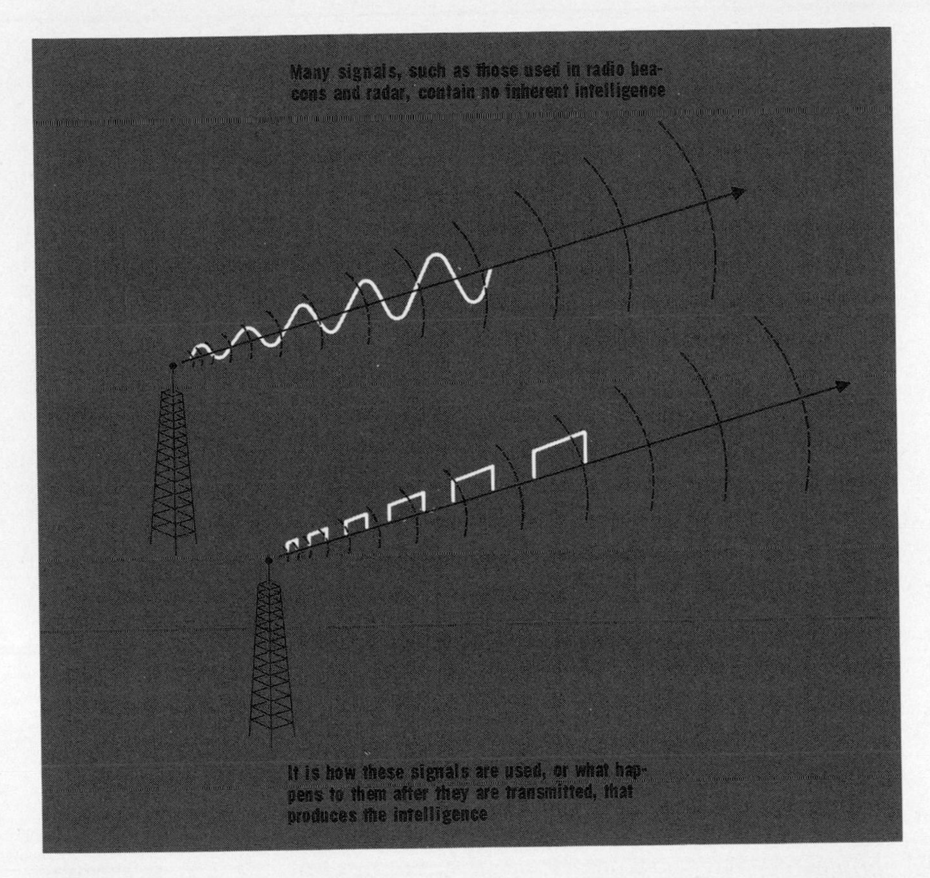

other signal intelligence

On the preceding pages, you have seen how electronic signals can modulate other waves for more efficient transmission of the desired intelligence. You have also seen how the complexity of signals can range from a simple continuous tone to a highly complex type having many components, such as the color television signal. In all cases, though, the electronic signals had some varying characteristic that represented intelligence. You will find that there are many applications involving electronic signals in which the signals themselves, as generated, contain *no* intelligence. Instead, intelligence is obtained by the way the signals are *used,* or by *interpreting* variations that occur after transmission. It might be said that these signals are *intelligence-providing* signals rather than intelligence-carrying signals. There are even cases where intelligence-carrying signals are used to provide information that is unrelated to the intelligence they are carrying.

Some examples of intelligence-providing signals are those used in radio direction finding, radio beacons, and radar. These as well as other typical types are described on the following pages.

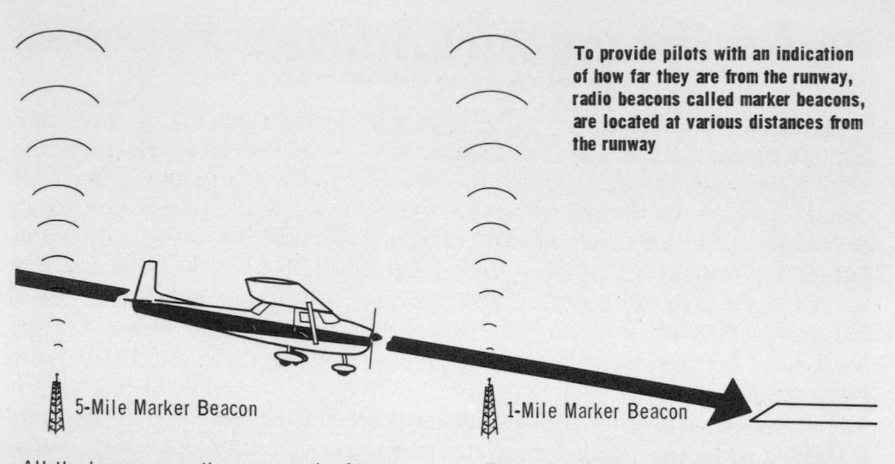

radio beacons

Radio beacons provide signals from *fixed locations* for use as navigational aids. Radio beacons contain no bearing or directional information themselves, but when they are used with *radio direction finding* equipment, they can provide very accurate information concerning bearing or location.

Generally, radio beacons are continuously *repetitive* signals. They may be an uninterrupted tone on a carrier, or either the tone or the modulated carrier may be interrupted to continuously repeat the dots and dashes of a single Morse code letter that identifies the particular radio beacon station. The basic characteristic of a radio beacon is that it is transmitted from a definite, fixed location, marked on appropriate navigation charts and maps. Furthermore, radio beacons are usually transmitted at *low power levels*. Thus, when an aircraft or ship radio picks up a radio beacon signal, it means that it is within some *maximum* distance from the known location of the radio beacon transmitter.

Certain types of radio beacons, especially during time of war, do not transmit unless they first receive a signal from the ship or place wanting their use. This is called *interrogating* the radio beacon. Usually, the interrogating signal is *coded* in some way so that only authorized planes or ships can make use of the radio beacon. When this type of system is used, the radio beacon must be controlled by a device called a *transponder*. This is a combination receiver/transmitter that receives the interrogating signal, and if it recognizes the signal as a legitimate one, triggers the radio beacon.

radio direction finding

Radio direction finding (RDF) is the process of using the radio signals themselves to determine the *relative direction* between the transmitter sending out the signals and the receiver picking them up. RDF should not be confused with the VOR navigation system previously described. VOR uses transmitting stations and signals specifically established for providing bearing information. RDF, on the other hand, makes use of *any* radio signal. For example, if a pilot was lost and tuned his RDF receiver to a station that he recognized as coming from St. Louis, he could easily determine his bearing from St. Louis with good accuracy.

The basic principle on which RDF works is that of the *directional* receiving antenna. This is an antenna whose ability to pick up radio waves varies, depending on its direction relative to the received waves. The characteristics of RDF antennas are such that if they are rotated a full 360 degrees, there are two points in their rotation where they pick up practically *no* radio waves. Both of these points occur when the plane of the antenna is perpendicular to direction in which the radio waves are traveling. This means, therefore, that if such an antenna is rotated until a received signal drops to its lowest or zero level, the plane of the antenna is head on to the direction of the station transmitting the signal.

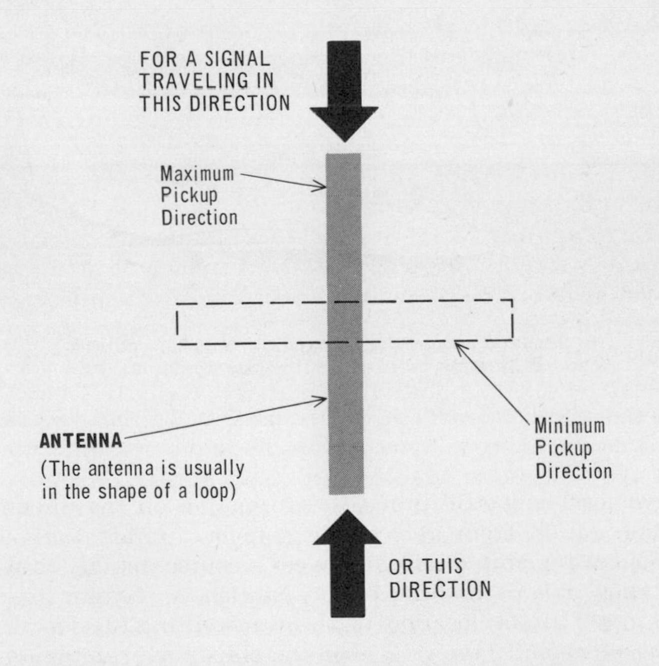

A directional antenna picks up practically no signal when its plane is in the same direction as that in which the signal is traveling

radio direction finding (cont.)

Basically, this is how RDF works: A signal is tuned in, and the antenna is turned until the signal practically or completely disappears. The plane of the antenna is then perpendicular to the direction of the transmitting station, and this direction is shown on some type of indicator. Actually, since the directional antenna picks up minimum signal in either of two opposite positions, the direction of the transmitting station might be in either of two *opposite* directions from the receiving antenna. The user of the RDF equipment can usually determine which of the directions is correct by using other navigation aids, such as a magnetic compass, and his knowledge of his approximate position with respect to the transmitting station. Some RDF equipment uses an auxiliary sensing antenna in conjunction with the rotating antenna. When this is done, only the one correct direction is indicated by the system.

By means of RDF, aircraft and ships can travel an accurate course directly towards any station they can receive; or they can travel in another direction using the station as a guidepost. This station might be a commercial broadcast station, or it might be a radio beacon.

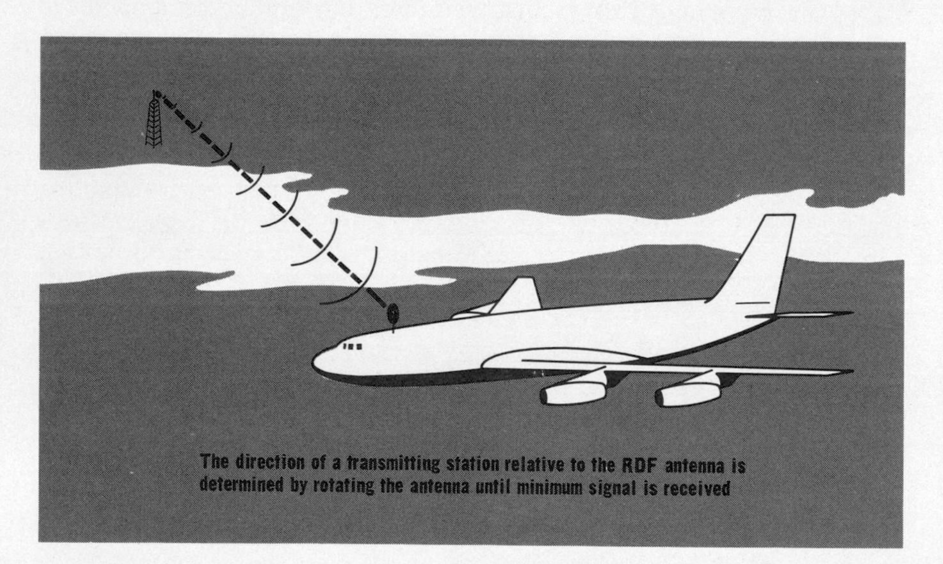

The direction of a transmitting station relative to the RDF antenna is determined by rotating the antenna until minimum signal is received

You have seen how RDF provides information on the direction of a transmitting station from a receiving ship or aircraft. Although this allows the receiving ship or aircraft to set a course directly to the transmitting station, it tells nothing of the *position* or location of the ship or aircraft, even if the location of the transmitting station is known. On a map, all it really shows is that the ship, for example, is located somewhere on a *straight line* that starts at the transmitting station and extends in the direction of the ship.

obtaining position information by rdf

If the direction of the ship from *two* known transmitting stations is known, the location of the ship can be determined. The result of each RDF measurement can be drawn as a straight line on a map, and the ship must be located where these two lines *cross*. This should be obvious, since the ship has to be somewhere on both lines, and the point of intersection is the only common point that the two lines have.

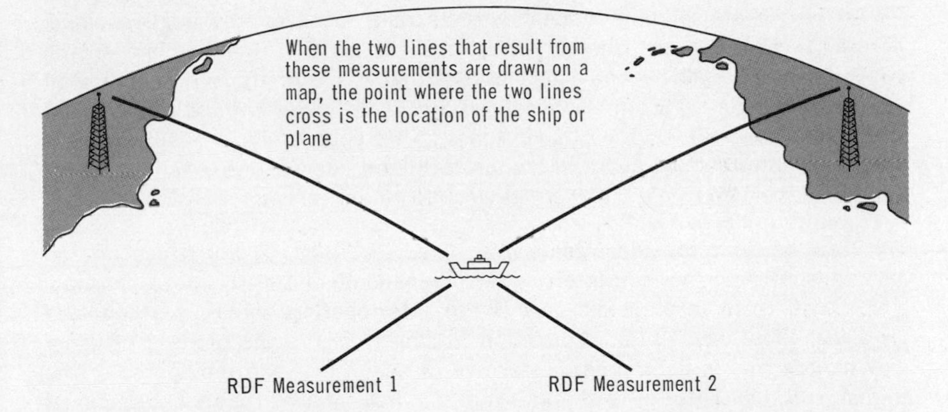

Since there is some degree of error involved in all RDF measurements, the location of the ship, as determined by the crossing lines, is also subject to some error. The degree of error can be reduced by taking RDF measurements on three transmitting stations. The crossing of three resulting lines then produces a small triangular area, and excellent accuracy results by assuming that the ship is at the center of this triangular area.

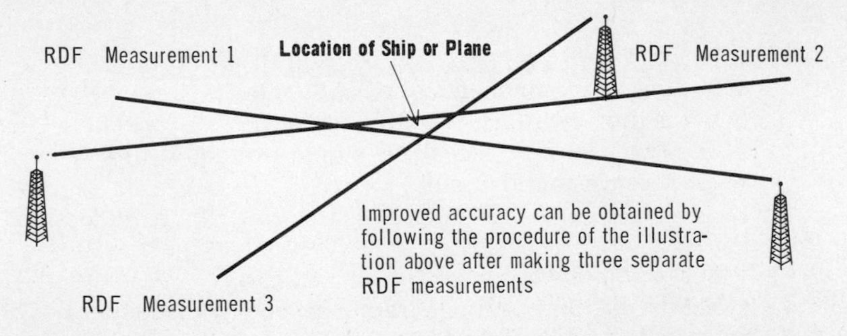

RDF measurements for fixing the location of a ship or plane can be taken with respect to any known transmitting station. Commercial broadcast stations, radio beacons, or even a combination of the two can be used.

summary

☐ The visual-aural range (VAR) system uses four separate signals, each transmitted on a separate carrier. Two of these are visual signals, and represent the letters N and A in Morse code. The other two signals are audible tones of 90 and 150 Hz. ☐ By means of VAR signals, a pilot can tell which quadrant he is in around the transmitting station. ☐ The visual omnirange (VOR) permits a pilot to read his exact bearing from a station on one instrument. ☐ VOR uses two signals: one is the reference phase and the other is the variable phase. The phase difference between the two indicates the bearing of the aircraft.

☐ An instrument landing system consists of a localizer and a glide slope. The localizer indicates whether an aircraft is to the right or left of the runway, and the glide slope indicates whether the aircraft is descending at the proper angle. ☐ Both the localizer and glide slope consist of a carrier modulated by a 90- and a 150-Hz tone. ☐ Both tones have equal strength when the aircraft is on course. If the aircraft is off course, one tone is stronger than the other, and this is presented visually on an indicator.

☐ Radio beacons provide signals from fixed locations. ☐ Some beacons do not transmit until they are interrogated. Transponders control these beacons, and cause them to transmit only if the interrogating signal is recognized as a legitimate one. ☐ Radio direction finding (RDF) is the process of using any radio signal to determine the relative direction between the source of the signal and the receiver picking it up. ☐ RDF makes use of directional antennas, whose ability to pick up radio waves depends on the direction of the antenna relative to the received waves. ☐ RDF can provide position information when two or more RDF measurements are made from separate transmitting stations.

review questions

1. Why can't VAR provide exact bearing information?
2. What are the components of a VAR signal?
3. What is the advantage of VOR over VAR?
4. If the two tones of a localizer signal are equal in strength, what does this tell the pilot?
5. How many carrier frequencies are used for a glide slope signal?
6. What is a *radio beacon*?
7. What is the purpose of a radio beacon transponder?
8. Why is a directional antenna required for RDF?
9. How can RDF provide position information?
10. Is AM or FM used for radio direction finding?

loran

Loran is a system of extremely accurate, long-range radio navigation for ocean-going ships and transoceanic aircraft. Its name is derived from its basic function: namely, *lo*ng *ra*nge *n*avigation. From the time of its initial development, various types of loran systems have come into use. One of these is called *standard loran,* and is used for commercial air and sea navigation. In this book, we will describe standard loran.

Loran operates on the basis of radio signals transmitted by a pair of shore-based stations. One of the stations is called the *master station* and the other the *slave station,* according to their function in the system. When a ship or aircraft receives the signals from both stations of the pair, it establishes its position as being somewhere on a line, called a *line of position,* that extends outward from the loran stations. As shown, a line of position is not straight, but shaped like a hyperbola. The exact location of the ship or plane can then be established by using signals from another pair of stations to determine a second line of position. The point where these two lines cross is the location of the ship or aircraft.

A loran line of position is determined on the basis of the *difference* in time required for a radic signal to reach the ship or aircraft from the master station and the slave station.

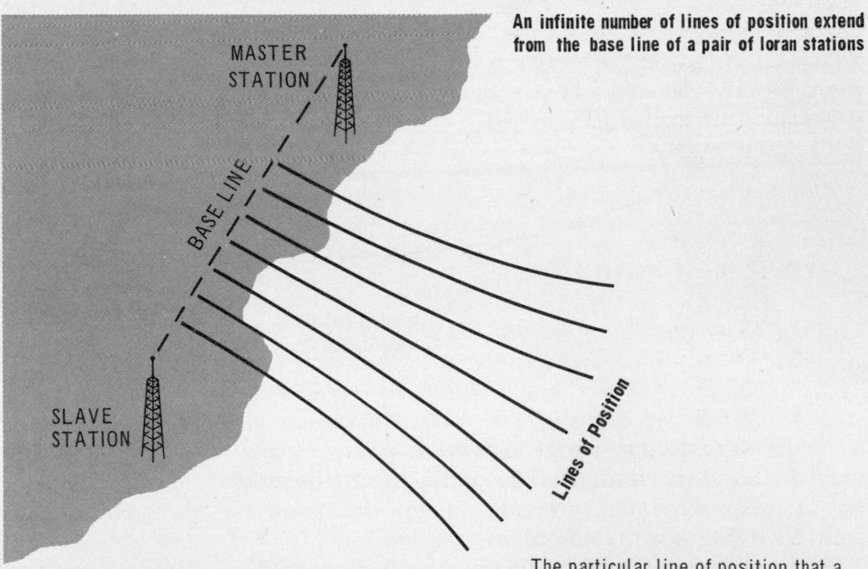

An infinite number of lines of position extend from the base line of a pair of loran stations

The base line is an imaginary line connecting the master and slave stations

The particular line of position that a ship is on depends on the exact difference in time that it takes a radio signal to reach it from the master station and slave station

loran (cont.)

The extreme accuracy of loran results because, for all practical purposes, radio waves travel at a *constant speed* of approximately 300,000,000 meters per second (186,000 miles per second). This makes it possible to convert the time that a radio wave travels directly into the distance traveled with a high degree of precision, since:

$$\text{Distance traveled} = \text{speed} \times \text{time}$$

So, if the speed of travel and the time traveled are known with accuracy, the distance can easily be found to the same degree of accuracy. As you will see later, this is the basis on which certain types of radar work. It is also fundamental to the operation of loran systems, although in loran it is the difference in the travel times of two waves that is important.

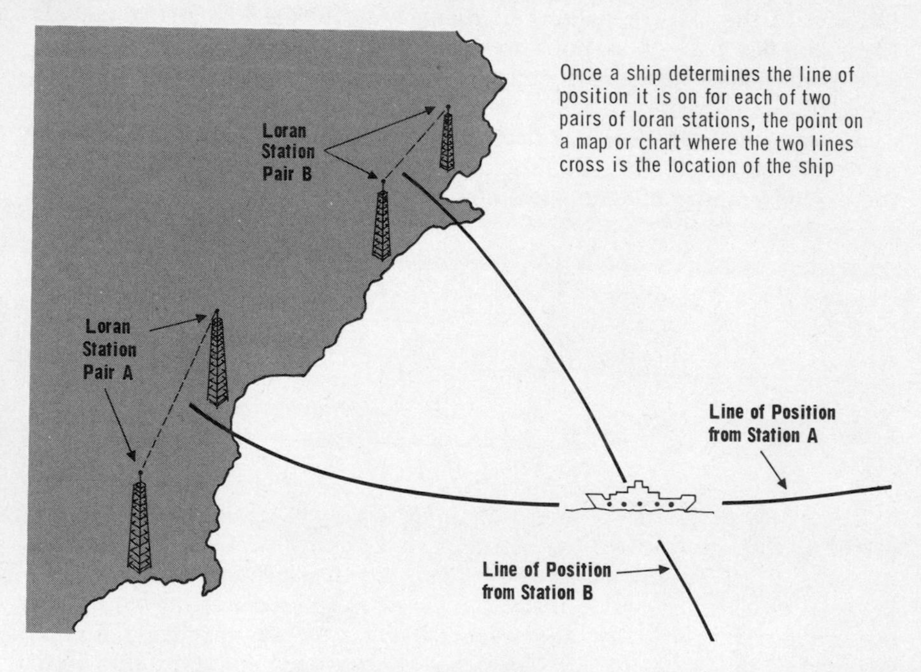

Loran signals consist of *repetitive pulses transmitted* by both the master and slave stations. The pulses from the master station are sent out at precisely timed intervals. The pulses from the slave station are controlled by those from the master station.

Each cycle of a loran transmission begins with a pulse transmitted by the master station. The direction of transmission is such that the pulse travels toward both the slave station and a ship, which we will assume is picking up the loran signals. The pulse arrives first at the slave station, and after a definite time delay causes that station to transmit its own pulse.

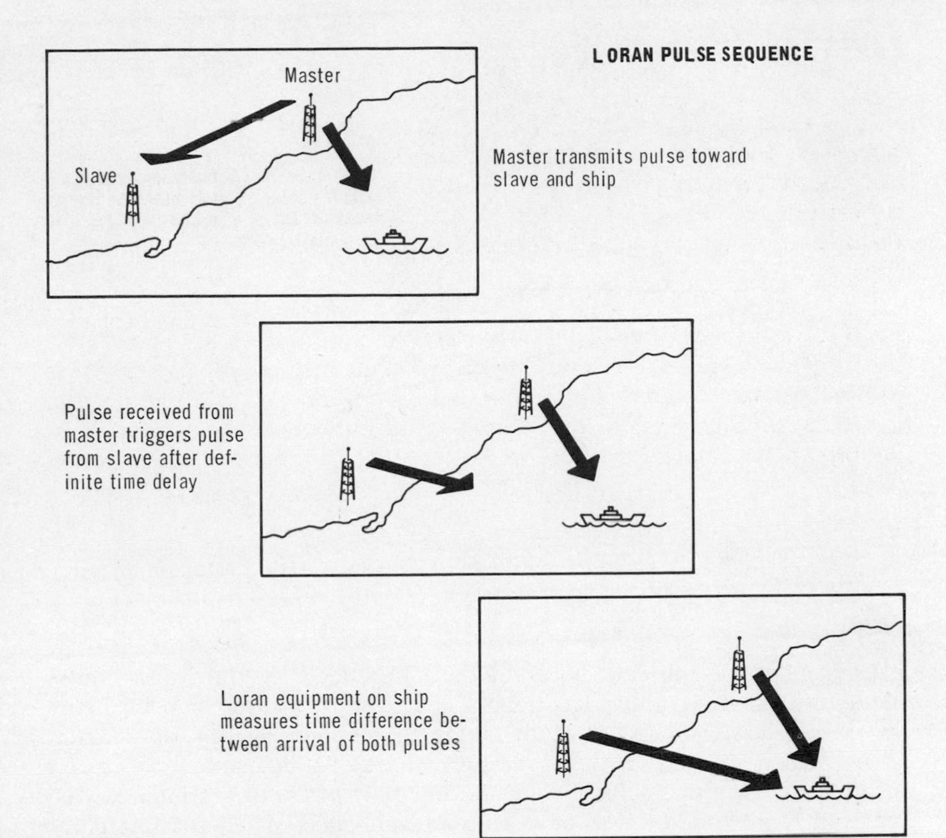

loran signals

The purpose of the time delay is to make the slave pulse always arrive at the ship during the second half of the time interval between master pulses. This is done to simplify the measurement of the time between master and slave pulses at the ship. Since the delay is always the same, it can be easily be compensated for in determining a line of position. Thus, there are *two* pulses traveling toward the ship: one from the slave, and the other from the master. The one from the master arrives at the ship first, followed by that from the slave.

Loran equipment on the ship picks up both pulses and measures the *time interval* between them. There are an *infinite* number of locations within the area served by the loran stations where the time interval between these two particular pulses would be the same. All of these locations, though, lie on *one* hyperbolic line, which is a line of position. Thus, by measuring the time between the slave and master pulses, the loran equipment establishes the line of position on which the ship is located.

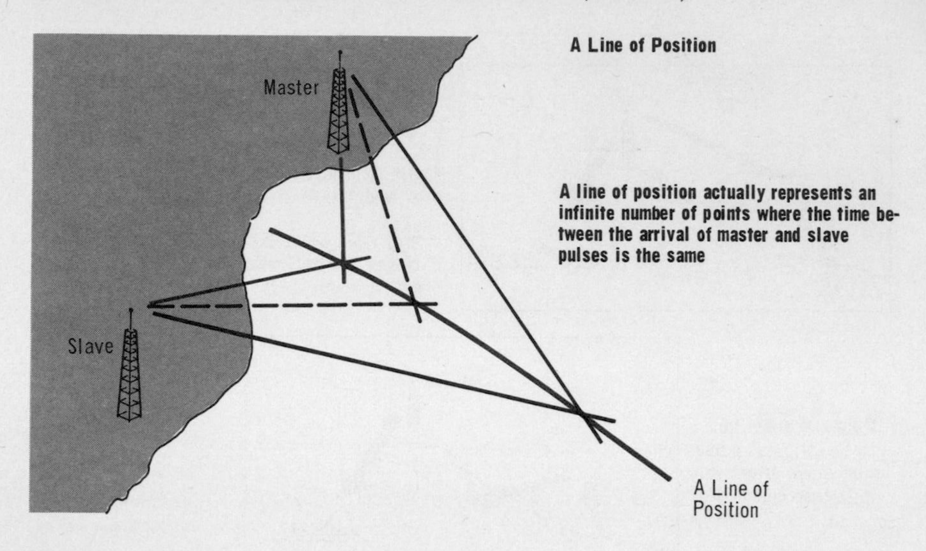

loran signals (cont.)

Conversion of the time interval data produced by the loran equipment into the geographic location of the line of position can be made on special loran charts, or from tables that allow the line of position to be drawn directly on a navigation chart. To determine the exact location of the ship, pulses from another pair of loran stations can be received, and another line of position established. The crossing point of the two lines then fixes the ship's location.

Loran stations operate continuously. So, the sequence of pulse transmission described is continually repeated.

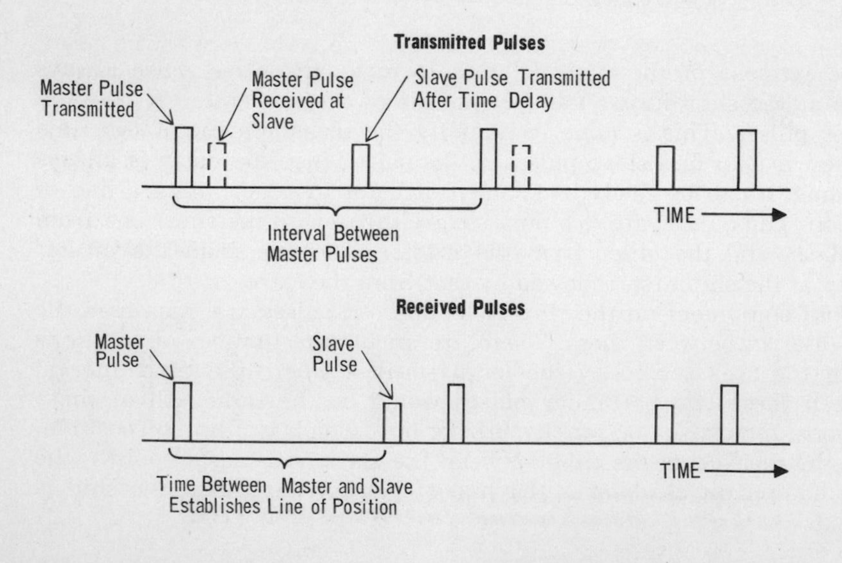

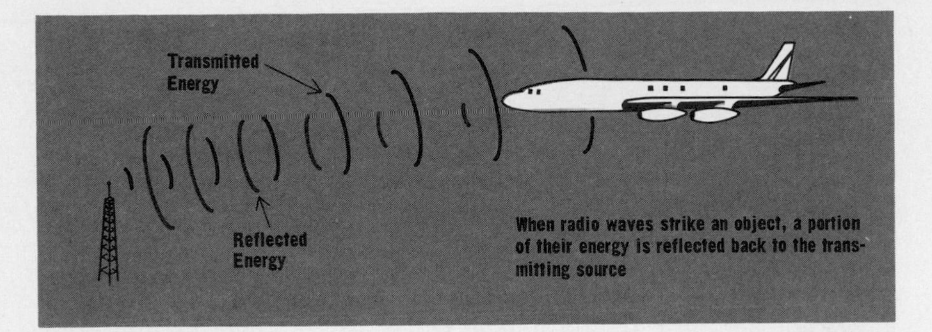

When radio waves strike an object, a portion of their energy is reflected back to the transmitting source

radar

Radar uses radio waves for detecting the *presence* and *location* of distant objects. The term "radar" was derived from *ra*dio *d*irection *a*nd *r*anging. Radar works because when a transmitted radio wave strikes an object, it is reradiated by the object back toward the radar antenna to indicate the presence of the reflecting object. If you can measure the *time lapse* between the pulse transmission and return of its *echo,* the distance of the reflecting object can be determined because the speed of the wave is considered constant at about 300,000,000 meters (186,000 miles) per second. So, by multiplying the speed by the total time lapse, the distance traveled is found. And, of course, the distance from the antenna to the reflecting object is one-half of the total distance.

Radar signals consist of narrow *pulses* of high-frequency waves. The intervals between pulses are considerably longer than the pulses themselves, and it is during these intervals that the reflected pulses are received. The radar circuits show the time lapse between a transmitted pulse and any echoes on a screen with visual displays calibrated in distance.

Radar antennas are *directional,* so they must be rotated to cover a 360-degree arc. This means that the *direction* of the reflecting object can also be obtained, since the antenna is facing the object at the instant the echo is received. The radar circuits are synchronized with the antenna, so that its bearing is also indicated on the visual display equipment with the echo.

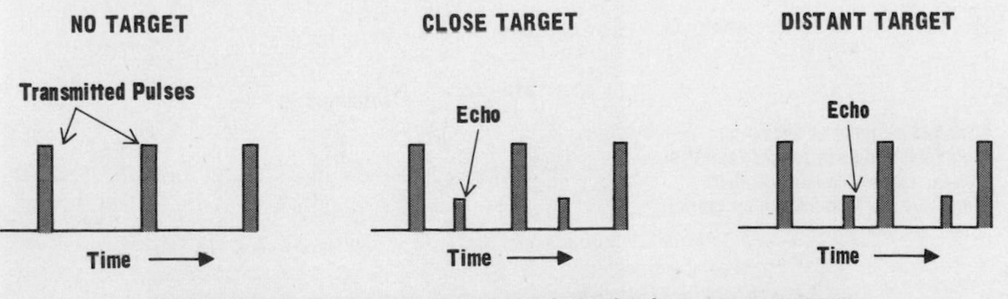

Since radio waves travel at a constant speed, the time lapse between a transmitted pulse and its echo is determined by the range of the reflecting object

characteristics of radar signals

Typical radar carrier frequencies in use today range from about 150 to 30,000 MHz. Pulse lengths can be anywhere from about 0.25 to 50 microseconds, with the exact value depending on the requirements of the particular system.

The time interval between pulses is determined by the *maximum distance* at which the radar is to be effective. The interval must be long enough to permit the echo of a pulse to return from a distant target *before* another pulse is transmitted. Otherwise, it would be impossible to associate an echo pulse with its corresponding transmitted pulse. And, if this were the case, the signal could not produce information concerning the range of the detected object. If the interval between pulses is too long, though, the rotating antenna might pass over the target during the interval, and fail to detect the target.

The number of pulses transmitted each second is called the pulse repetition rate (PRR) of the signal. The PRR is determined by the pulse width and the time interval between pulses

An important characteristic of radar signals is the ratio of their peak to average, or effective, power. Since the power in a reflected echo is a very small percentage of the power in a transmitted pulse, it is necessary that the transmitted pulse contain considerable power. The pulses of actual radar signals, therefore, often contain many millions of watts of power. As you will recall, though, the average power of a pulse signal is much less than the peak power of the individual pulses. As a result, even though radar transmitters put out a great amount of usable power in the pulses, the average power they supply is much lower.

The radar signal is shown as rectangular pulses for convenience Actual signals consist of short bursts of the high-frequency carrier

doppler signals

A number of electronic equipment used for detecting objects and measuring their distance and velocity make use of *Doppler signals*. These signals are named after the Austrian physicist Christian Doppler, who discovered the phenomenon of the Doppler effect. The Doppler effect is common to all physical *wave motion*, and you are probably familiar with it, even though you may not be aware of its scientific explanation.

Basically, the Doppler effect means that the *frequency* of a wave emitted by a source is not necessarily the frequency picked up by a receiver, where a receiver here means either a person listening to a sound wave, or an antenna picking up a radio wave. The frequency at the receiver depends on the *relative motion* between the source and receiver. If both the source and receiver are *stationary*, the frequency at both is the *same*. But, if either one, or both, is moving *away* from the other, the frequency at the receiver is *lower* than that emitted by the source. Conversely, if the receiver and source are moving *closer together*, the frequency at the receiver is *higher* than that emitted by the source. A common example of the Doppler effect is the pitch of a train whistle as heard by an observer standing alongside the tracks. As the train approaches, the whistle has a certain pitch. But when the train passes the observer, there is a sudden drop in the pitch of the whistle. This is caused by the change in relative motion between the train and the observer.

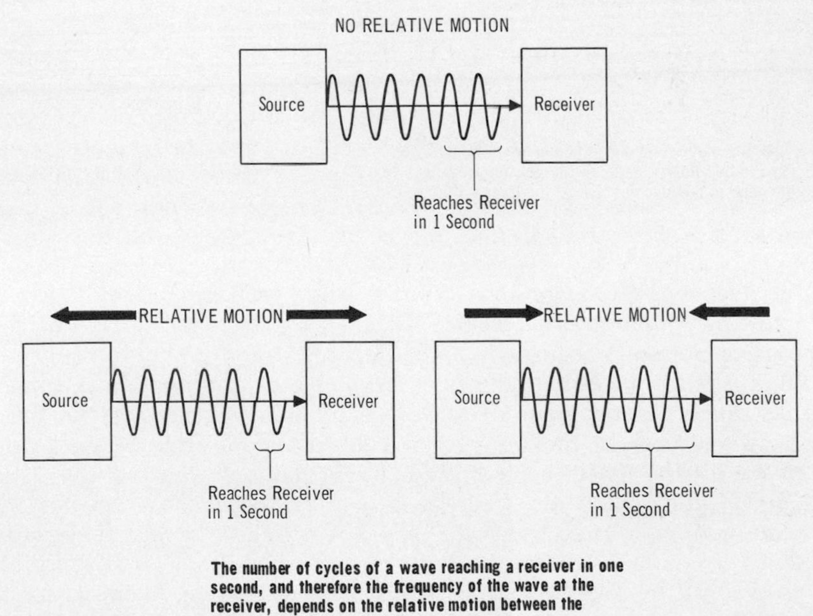

The number of cycles of a wave reaching a receiver in one second, and therefore the frequency of the wave at the receiver, depends on the relative motion between the receiver and the source emitting the wave

doppler radar

The Doppler effect on radio waves is easily understood if you consider the frequency at a radio receiver as the number of cycles of the wave that *reaches* the receiver each second. With both the source and receiver stationary, the radio waves travels toward the receiver at the speed of light. When the distance between the source and receiver is increasing, though, it is as if the wave slowed down with respect to the receiver; so fewer cycles reach the receiver each second. The converse is true when the distance between the source and receiver is decreasing.

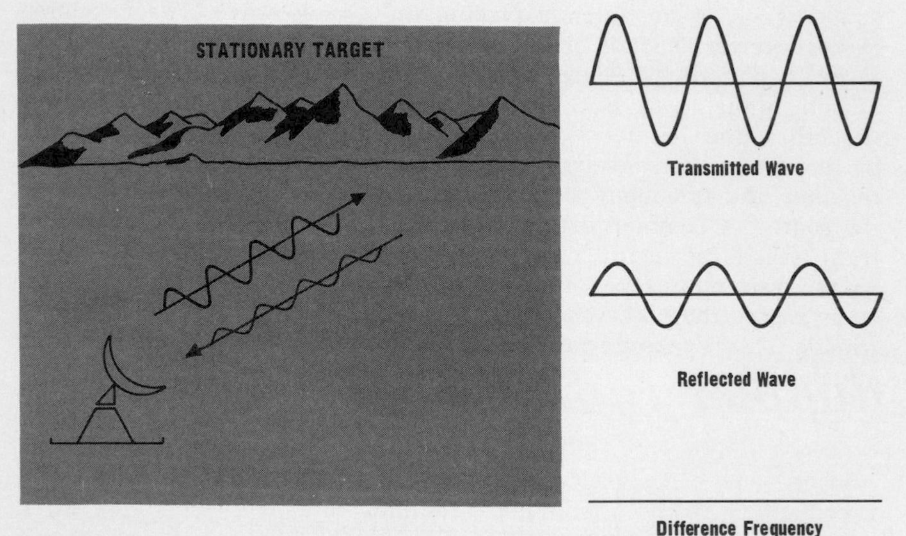

When the signal from a continuous-wave Doppler radar is reflected from a stationary target, the frequency of the echo is the same as that of the transmitted wave

Since the two waves have the same frequency, mixing them produces no signal at the difference frequency

The frequency change that results from relative motion between an emitting source and a receiver is called the *Doppler shift*. The fact that a Doppler shift occurs only when there is relative motion is put to use in a type of radar that detects *moving objects* and ignores *stationary objects*. Such radar is called *Doppler radar*, and is extremely useful in applications where a moving target, such as an aircraft, must be distinguished from fixed objects that might be received by the radar antenna at the same time. There is a definite relationship between the frequency of an emitted wave, the speed at which it travels, the relative motion between source and receiver, and the Doppler shift that occurs. As a result of the relationship, it is possible not only to detect a moving object, but to measure its velocity.

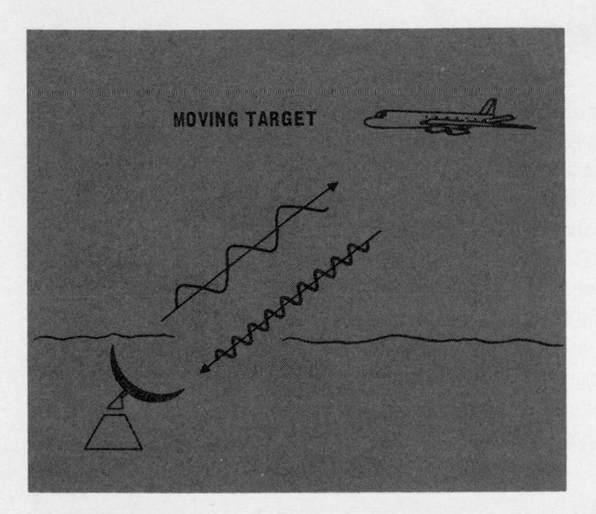

When the signal is reflected from a moving target, the echo undergoes a Doppler shift. When the target is approaching the antenna, the echo is shifted to a higher frequency

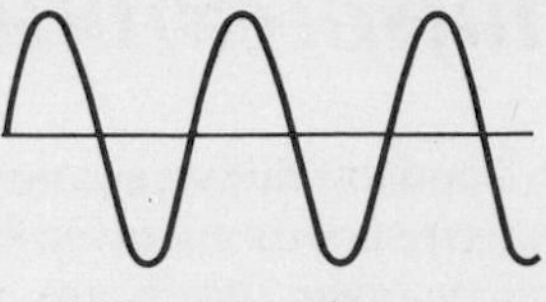

Reflected Wave

Difference Frequency

Since the two waves have different frequencies, mixing produces a signal at the difference frequency

doppler radar (*cont.*)

In one type of Doppler radar, a *continuous wave* is transmitted in a directional beam by a scanning (rotating) antenna. When the wave strikes an object, a portion of the wave is reflected toward a receiving antenna, which picks up the reflected wave and transfers it to receiver circuits. These receiver circuits also continuously receive a portion of the transmitted wave directly from the transmitting circuits.

In the receiving circuits, the reflected wave is mixed with the transmitted wave in a process, which like amplitude modulation, produces sum and difference frequencies. For this application, it is the *difference frequency* that is significant. When the reflected wave is from a stationary object, it has the same frequency as the transmitted wave, so the difference frequency produced by the mixing process is zero. But, when the reflected wave is from a moving object, it undergoes a Doppler shift as a result of the relative motion between the detected object and the receiving antenna. Thus, when this wave is mixed with the transmitted wave, there is a difference frequency produced. The *existence* of the difference frequency indicates the *presence* of the detected object. The exact *magnitude* of the difference indicates the *relative velocity* between the object and the receiving antenna. With this type of Doppler radar signal, though, the *range* of the detected signal cannot be determined.

pulse doppler radar

Doppler radar signals can be made to produce *range* information by transmitting *pulses* instead of a continuous wave. When the echoes are received, they are processed differently from conventional radar. Effectively, each reflected pulse is *subtracted* from an exact *duplicate* of its corresponding transmitted pulse. This subtraction is an algebraic subtraction of the instantaneous pulse amplitudes and its results are different for reflected pulses from stationary and moving objects.

When reflected pulses from a stationary object are subtracted from their corresponding transmitted pulses, the result is pulses all having the same *constant* amplitude. This occurs regardless of the phase of the reflected pulses. Of course, echoes from different objects will have different phases. But all the pulses from any one object will have the same phase, so the pulses that result from the subtraction process will have a constant amplitude.

Reflected pulses from a moving object have *different* phases, with the phase of each succeeding pulse changing as the object moves closer to or farther away from the radar antenna. This difference in phase causes the pulses that result from the subtraction process to *change* in amplitude. The *rate* at which these pulses change in amplitude corresponds to the Doppler shift caused by the motion of the detected object. Thus, a moving target is indicated by varying-amplitude pulses from the subtraction process. The velocity of the detected object is indicated by the rate at which these pulses change in amplitude. And the range of the object is indicated, as in conventional pulse radar, by the time interval between a reflected pulse and its corresponding transmitted pulse. Radar that operates in this way is often called *moving-target-indicator* (MTI) radar.

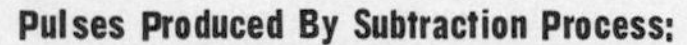

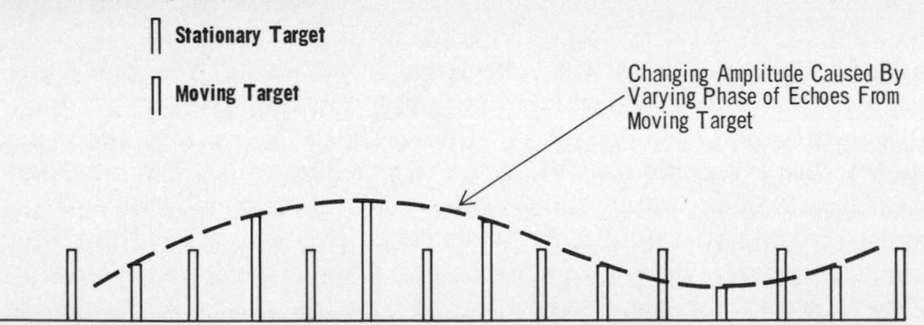

MTI radar differentiates between the stationary and moving targets by subtracting echo pulses from exact replicas of their corresponding transmitted pulses

This subtraction produces constant-amplitude pulses for stationary targets, and varying-amplitude pulses for moving targets

sonar

Sonar is used for detecting objects *under water* and measuring their range, and so is similar to radar. However, radar uses signals having extremely high *radio* frequencies; sonar signals are actually *physical sound waves* in or near the *supersonic band*. The reason for this is that radio waves cannot be sent under water, and high-frequency sound signals are greatly attenuated after traveling only short distances. So, to achieve a satisfactory transmission range, much lower frequencies are used. Sonar means *so*und *na*vigation and *r*anging.

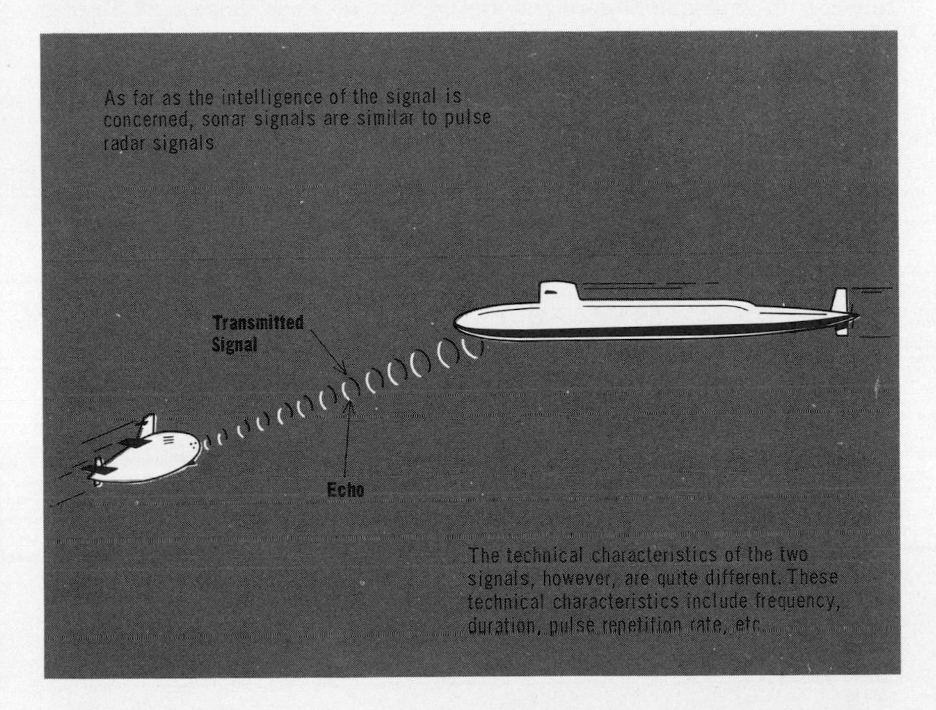

Sonar signals consist of *pulses* of a fixed frequency that are transmitted into the water by a powerful device that works similar to a loudspeaker. The pulses are usually a few seconds in duration, and have a frequency somewhere between about 10 and 50 kHz. During the interval between transmitted pulses, the sound transmitter functions as a microphone to pick up any echoes. Since the speed of sound in water is a *known quantity*, the time interval between a transmitted pulse and its echo allows the range of the detected object to be determined. The transmitter-microphone is directional, making it possible to determine the direction of the object as well as its range. Sonar signals are also affected by the Doppler effect, so the pitch of the echo will change if the target is moving toward or away from you.

Sonar signals are also used for measuring *depth*, and to locate schools of fish.

radar altimeters

Electronic signals are frequently used for the measurement of height. The most common example is the radio altimeters used in aircraft for determining the altitude of the aircraft above ground. Basically, there are two broad types of radio altimeters: one that employs radar principles, and another that uses FM signals.

Radar altimeters are essentially pulse radars used to determine range, which in this case is the distance between the aircraft and the ground. Pulses are transmitted downward from the aircraft, and the time interval between their transmission and the return of their echoes is directly proportional to the altitude. Circuits in the altimeter convert the measured time intervals into distance indications for visual presentation.

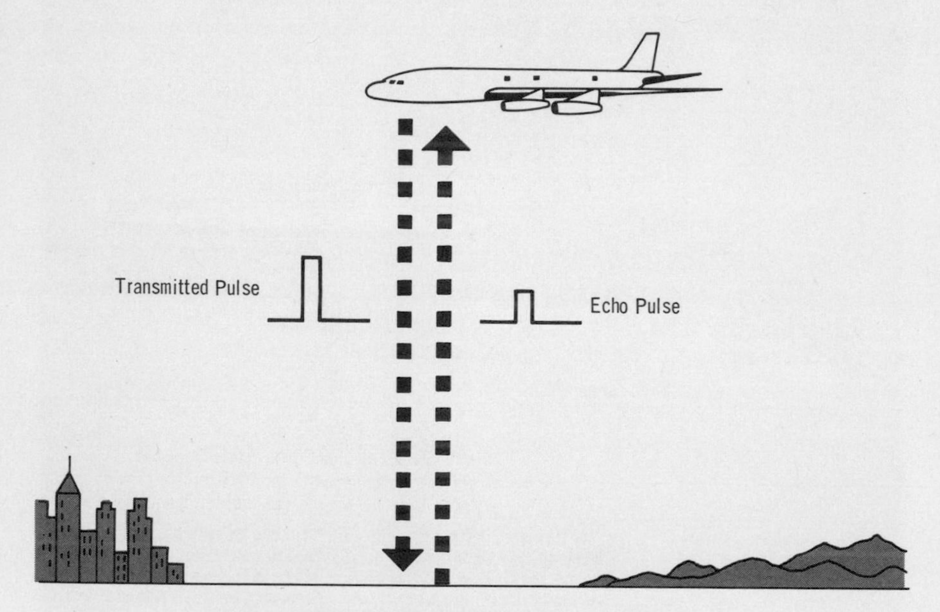

In radar altimeters, the time interval between a transmitted pulse and its echo is converted directly into altitude

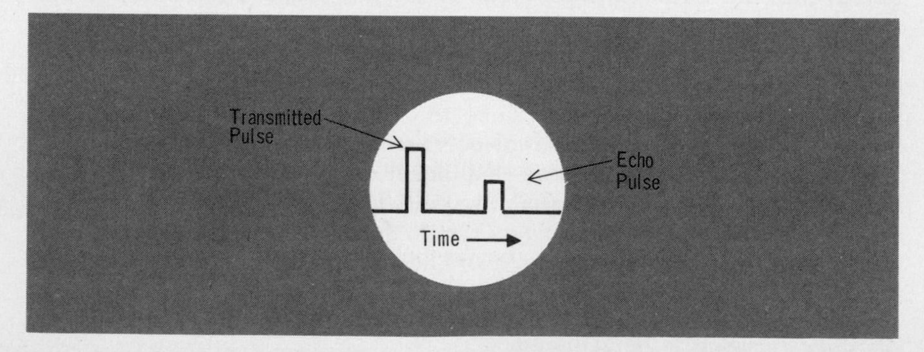

FM altimeters

FM altimeters also transmit signals toward the ground and use their return echoes to measure the altitude. However, it is the frequency of the returning echoes that is used to determine the altitude. This is possible, since the pulses consist of an FM carrier that has been modulated by a *triangular* pulse. You recall from your reading of FM that in this type of signal the frequency of the carrier increases linearly from its center frequency, and then abruptly starts to decrease linearly until it returns to its center frequency. By the consideration of just the increasing portion of the pulse, this means that in equal time intervals, the frequency increases by equal amounts. Thus, there is a direct relationship between the instantaneous frequency at any point in the signal and the time from the beginning of the pulse to that point.

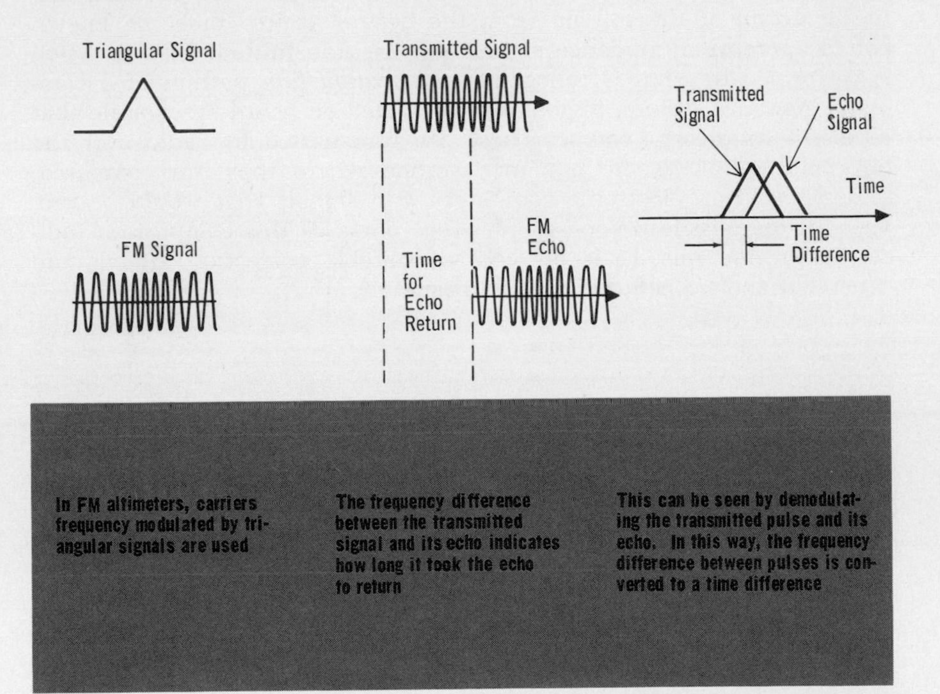

In FM altimeters, carriers frequency modulated by triangular signals are used

The frequency difference between the transmitted signal and its echo indicates how long it took the echo to return

This can be seen by demodulating the transmitted pulse and its echo. In this way, the frequency difference between pulses is converted to a time difference

When the echo of a signal is received, its instantaneous frequency is compared with that of the transmitted signal, which is actually still being transmitted. The difference in their frequencies then indicates the time that has elapsed from the instant transmission of the signal began to the instant its echo began being received. In other words, by measuring the *frequency difference* between the transmitted and reflected signals, the *time* of travel of the signal, and, therefore, the distance traveled is determined.

missile guidance

Electronic signals are widely used in the field of missile guidance. By means of these signals, the missile is controlled by a source *external* to the missile. This is called *command guidance.* Normally, the function of guidance signals is to control the operation of auxiliary rocket motors, control surfaces, or other controlling devices and thereby alter the direction of the missile to put it on the right course. There is an almost limitless variety of command guidance signals, ranging from a simple sequence of pulses or tones, to extremely complex pulse and subcarrier modulation systems. In a simple system, one tone or pulse code can mean "turn right," and another tone or pulse code can mean "turn left." This can be done for "climb," "dive," "go faster," "go slower," "explode," and so on.

Since guidance signals are used to correct the course of a missile, the *deviation* of the missile from the desired course must be known before appropriate guidance signals can be transmitted. This function is performed by what is called the *information link* portion of a command guidance system. Signals are generated on board the missile that represent its present course. These are transmitted by radio over the information link to the guidance station where they are compared, sometimes in an electronic computer, with signals that would be produced if the missile was on the desired course. If this comparison indicates that the missile is off course, suitable correction signals are generated and transmitted to the missile.

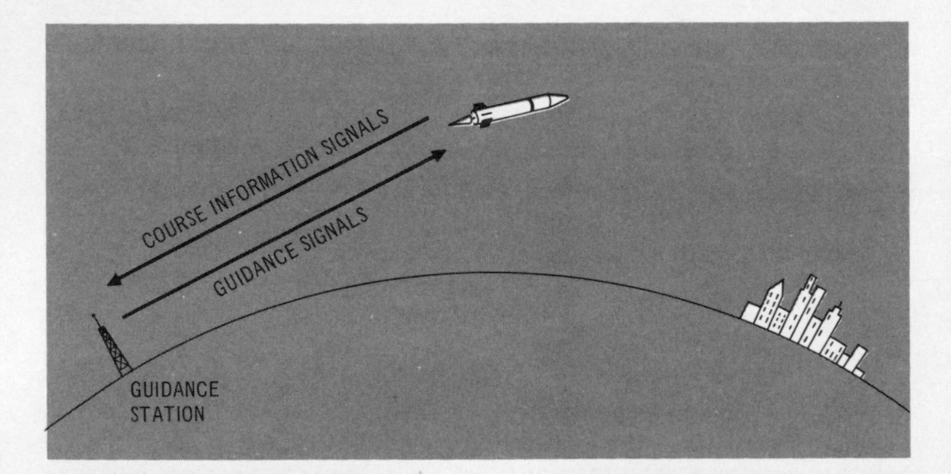

Missile command guidance signals are of two types: those that carry information on the actual course of the missile, and those that carry the correction information used by the missile to change direction to the desired course

Both types of these signals are part of any command guidance system. In some systems, the correction information is generated by the missile itself

missile guidance (cont.)

In another type of command guidance system, the actual course of the missile is determined by *radar tracking stations.* This course is then compared, again usually by an electronic computer, with the desired course, and appropriate course-correction signals are generated and transmitted to the missile by radio.

You should not get the impression from the above discussion that all missiles are guided by *radio signals.* There are various systems of missile guidance in which all of the guidance equipment is located right *on board* the missile. Signals are still generated that indicate the actual course of the missile; and these signals are used to produce guidance signals. However, all of this is done on board the missile.

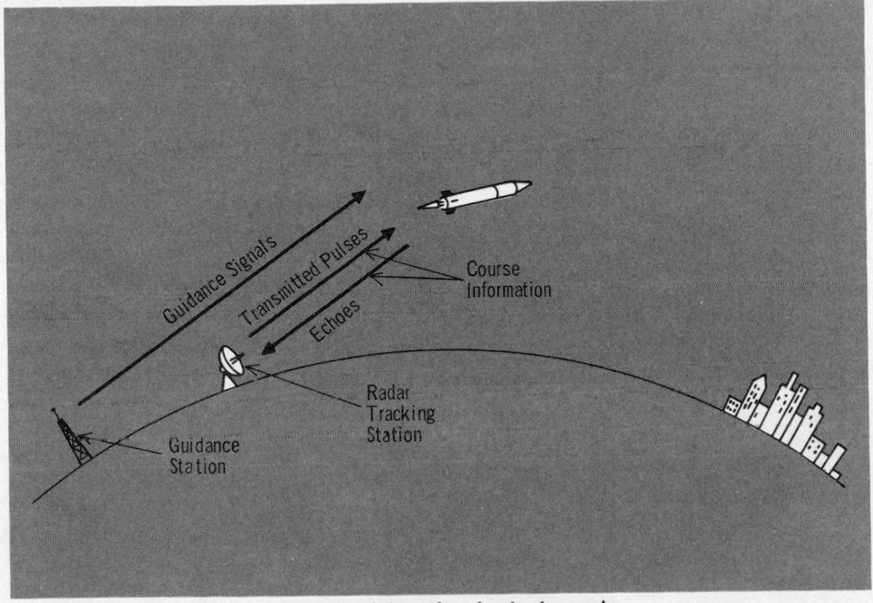

The course-information signals also can be produced by radar tracking stations

Missile homing systems consist of equipment on board a missile that permits the missile to seek a target on its *own,* without the need for external command signals. In these systems, the missile follows, or *homes in* on, a signal emitted by the target. There are three basic types of homing systems, grouped according to the *original* source of the target signal that they home in on. They are *active, semiactive,* and *passive* systems.

Active homing systems generate a signal, transmit it toward the target, and then home in on the reflection, or echo, of this signal. Radar signals are used extensively for such active systems, since they allow both the *distance* and *direction* of the target to be determined.

missile homing

In semiactive systems, the missile also homes in on reflected signals from the target. However, the missile itself does not transmit the original signal to the target. Instead, it is sent to the target from some *other* source, such as a ground station, the aircraft firing the missile, etc.

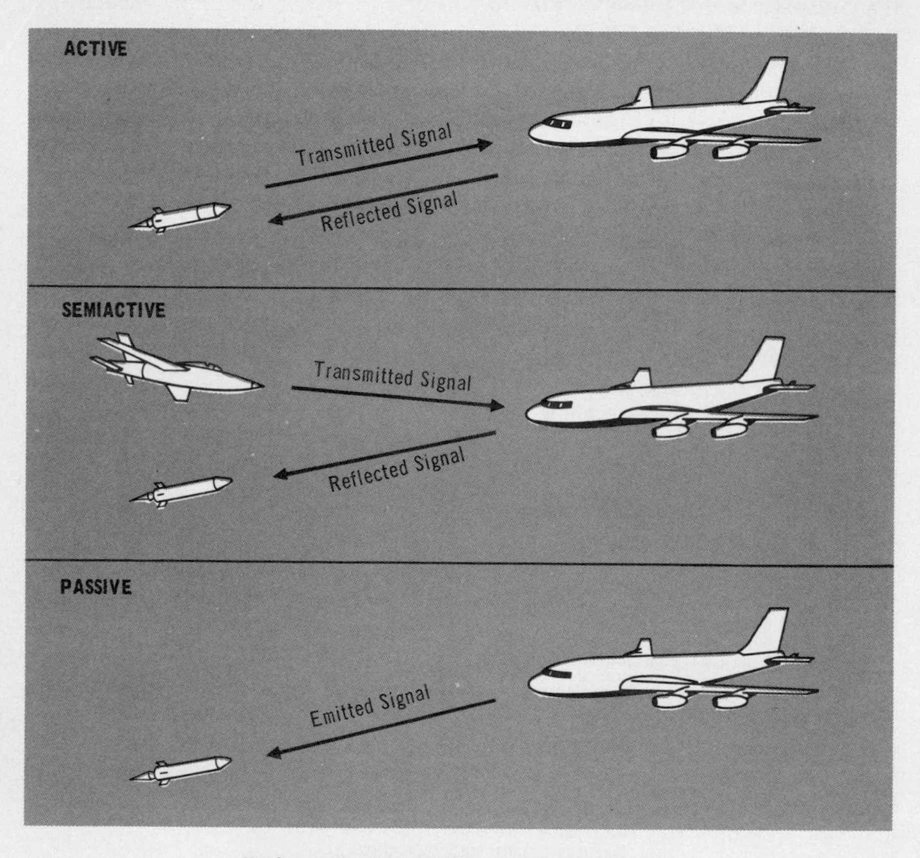

Homing systems are classified according to the original source of the homing signal

In passive systems, the missile homes in on some form of radiation emitted *directly* by the target. Usually, this radiation is not electronic, but in the form of light, heat, or sound waves. There are other even more sophisticated forms of missile guidance, such as beacons and map readers. In beacon guidance, a manned plane can fly over enemy territory and drop a number of radio beacons at many strategic locations, and then the signals from the beacons will provide a homing beam for the missiles. A map-reading missile can make a radar map of its course over land to check its course and keep itself on target.

summary

□ Loran is a very accurate long-range radio navigation system. □ Loran signals received from a master station and a slave station establish the location of a ship or plane somewhere on a line of position. Signals received from two pairs of Loran stations then locate the exact position at the point where the two lines of position cross. □ Radar uses radio waves for detecting the presence and location of distant objects. □ Radar works by measuring the time lapse between the transmission of a pulse and the return of its echo, or reflection, from a target. □ Directional antennas are used to provide information on the bearing of the target relative to the radar transmitter.

□ A Doppler shift occurs when there is relative motion between a source emitting radio waves and a receiver. □ Doppler radar makes use of the Doppler shift to detect moving objects. It does so by comparing the frequency of the transmitted wave with the frequency of the reflected wave. □ Doppler radar can produce range information, as well as detection, if pulses are transmitted instead of a continuous wave. □ Sonar is a method of detecting objects under water and measuring their range. It does this by transmitting sound pulses and detecting the echoes returned from targets.

□ Radar altimeters are essentially pulse radars that measure the distance between an aircraft and the ground. □ FM altimeters make use of triangular pulses that frequency modulate a carrier. Distance is determined by measuring the frequency difference between the transmitted and reflected signals. □ Command guidance uses electronic signals from an external source to control the path of a missile. □ Missile homing systems can be classified as active, semiactive, or passive, depending on the original source of the signal that they home in on.

review questions

1. Why does Loran require both a master and a slave station?
2. What is a radar echo?
3. Why are directional antennas required for radar?
4. In pulse radar, what determines the time interval between pulses?
5. What is the *Doppler shift*?
6. Can Doppler radar detect stationary objects?
7. How can Doppler radar be made to produce range information?
8. Why are radio waves not used for detecting objects under water?
9. How does an FM altimeter determine height?
10. What is *command guidance*?

facsimile

Facsimile is a process by which electronic signals are used to transmit *visual material,* such as still pictures, printed matter, and maps. The material to be transmitted is placed on a *revolving drum* and scanned by a narrow beam of light so that every spot on the material is touched by the light beam in some sequence.

During the scanning, the light striking the picture is reflected to a photocell, which produces a signal voltage proportional to the *intensity of the reflected light.* The signal modulates an r-f carrier for transmission.

Production of Facsimile Signal

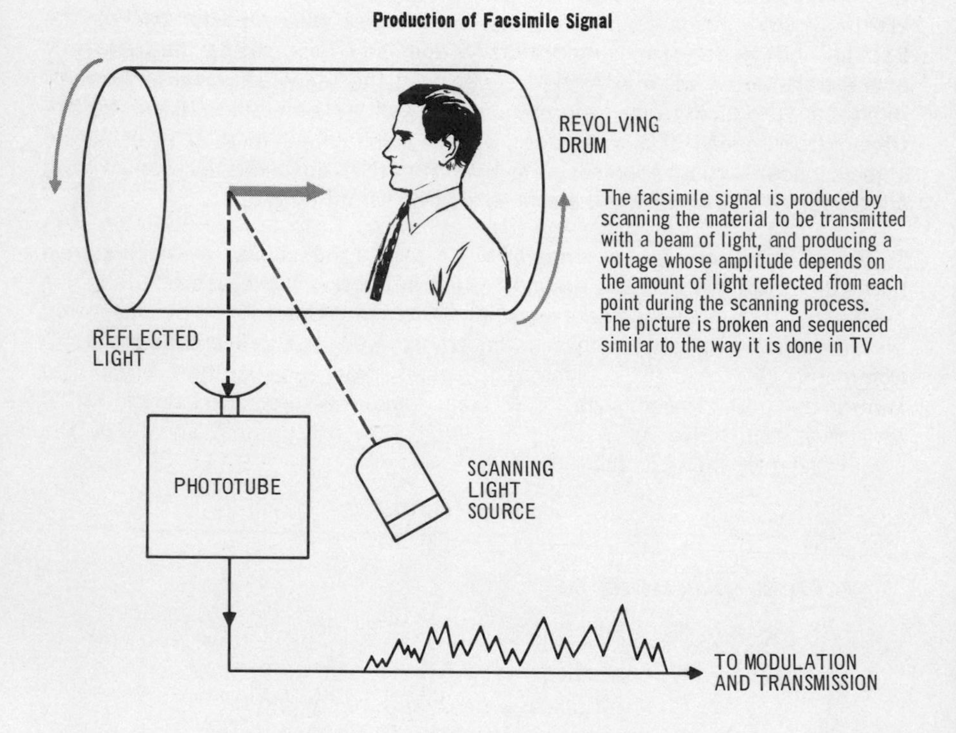

The facsimile signal is produced by scanning the material to be transmitted with a beam of light, and producing a voltage whose amplitude depends on the amount of light reflected from each point during the scanning process. The picture is broken and sequenced similar to the way it is done in TV

After transmission, the signal is demodulated and reproduced in various ways. One is to use *photographic paper* that is exposed to a special gas-filled lamp whose light intensity can be varied by the signal. So when the photographic paper is developed, an image of the original material appears.

Another method is to use special paper that allows electric current to pass through it, and turns dark at that spot, in proportion to the signal current. A pen-like *stylus* scans the paper to provide the current.

At the transmitter, synchronizing signals are added to the carrier so that the receiving equipment follows the same scanning pattern. This is similar to what is done in television.

FCC frequency bands

As has been stated, the Federal Communications Commission allocates frequency bands for all radio transmission within the United States. The need for government regulation of radio transmission is obvious once you realize the chaos that would result if anyone could transmit radio waves, and do so at any power level and frequency. The intelligence carried by many of the waves would become unintelligible and therefore useless as a result of the interaction and interference that would occur.

To prevent this interference, the FCC divides the radio-frequency spectrum into various bands, and limits the use of these bands to certain types of radio transmissions. The FCC also regulates the use of the particular frequencies in any one band to prevent interference between signals transmitted within the *same* geographic area. In addition to its regulation of frequencies, the FCC also limits the amount of power that can be used to transmit the various types of signals. This prevents signals from carrying beyond their intended area of coverage and interfering with radio reception in distant geographic areas.

Many of the frequency assignments made by the FCC are listed on this and the following page. A complete list can be found in standard handbooks or other reference works, or obtained from the FCC.

FCC FREQUENCY BANDS

Band	Frequency Range	
Standard Broadcast Band (540 to 1600 kHz): It is divided into 107 channels, each having a 10-kHz bandwidth.	540 kHz 550 kHz 560 kHz 570 kHz 580 kHz	590 kHz 600 kHz 610 kHz And so on to 1600 kHz
Commercial FM Broadcast Band (92.1 to 107.9 MHz): It is divided into 80 channels, each having a 200-kHz (0.2-MHz) bandwidth.	92.1 MHz 92.3 MHz 92.5 MHz 92.7 MHz 92.9 MHz	93.1 MHz 93.3 MHz 93.5 MHz And so on to 107.9 MHz
Television Channels: There are 82 channels, each having a 6-MHz bandwidth. What was channel 1 is now, for the most part, taken up by amateur radio operators.	Low VHF Band: 54-60 MHz (channel 2) 60-66 MHz (channel 3) 66-72 MHz (channel 4) 76-82 MHz (channel 5) 82-88 MHz (channel 6) High VHF Band: 174-180 MHz (channel 7) 180-186 MHz (channel 8) 186-192 MHz (channel 9) 192-198 MHz (channel 10) 198-204 MHz (channel 11) 204-210 MHz (channel 12) 210-216 MHz (channel 13) UHF Band: 470-476 MHz (channel 14) 476-482 MHz (channel 15) And so on to channel 83 at 884-890 MHz	

FCC frequency bands (cont.)

Many of the following frequency bands are subject to frequent change by the Federal Communications Commission.

FCC FREQUENCY BANDS		
Band	**Frequency Range**	
Radio Navigation (Nonaeronautical)	10-14 kHz 90-110 kHz 1800-2000 kHz 2900-3300 MHz	5250-5650 MHz 8500-9800 MHz And others
Aeronautical Radio Navigation	200-285 kHz 325-405 kHz 1605-1800 kHz 108-118 MHz 328.6-335.4 MHz	960-1215 MHz 1300-1660 MHz 2700-3300 MHz And others
Aircraft (Air-to-Ground) Communications	325-405 kHz 2850-3155 kHz 3400-3500 kHz 6525-6765 kHz 23.2-23.5 MHz	118-132 MHz 6425-6575 MHz 16,000-18,000 MHz And others
For Government Use	24.99-25.01 MHz 27.54-28.00 MHz 34.00-35.00 MHz 132.00-144.00 MHz 148.00-152.00 MHz	157.05-157.25 MHz 4400-5000 MHz 13,225-16,000 MHz And others
For Land Transportation (Taxicabs, railroads, etc.)	30.64-31.16 MHz 43.68-44.60 MHz 72.00-76.00 MHz 452-453 MHz 890-940 MHz	2110-2200 MHz 6425-6875 MHz 11,700-12,700 MHz And others
Amateur Radio	3500-4000 kHz 7100-7300 kHz 28.00-29.70 MHz 50.00-54.00 MHz 220-225 MHz	2300-2450 MHz 5650-5925 MHz 10,000-10,500 MHz And others

special signal considerations

You have now seen how electronic signals can be made to both carry and produce a wide variety of information. You have also seen how although signals possess various properties, the intelligence being carried is usually only contained in *certain* of these properties. Any changes in these *intelligence-carrying properties* that occur during transmission or processing of the signal cause distortion of the signal intelligence, and are therefore undesirable. Other properties of a signal, though, which do not represent intelligence, can be changed or distorted without affecting the usability of the signal in any way. And, as a matter of fact, in many practical applications certain characteristics of signals are changed without distorting their intelligence so that the signal may be processed in the most effective and efficient manner. Examples of this include changing the frequency of the carrier on which a signal is carried, as well as changing the shape of the individual pulses of certain pulse signals.

Of course, since signals can represent intelligence in so many ways, it is impossible to make general statements about what can and what cannot be done to a signal without distorting it to the point where its usability is impaired. This can only be done by having a good understanding of the nature of the signal, how it carries its intelligence, and the degree of distortion of the intelligence that is permissible.

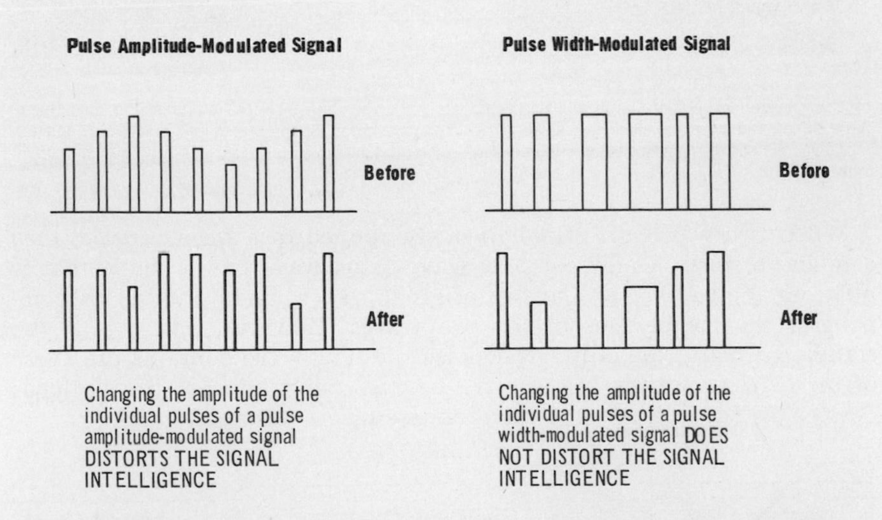

Changing the amplitude of the individual pulses of a pulse amplitude-modulated signal DISTORTS THE SIGNAL INTELLIGENCE

Changing the amplitude of the individual pulses of a pulse width-modulated signal DOES NOT DISTORT THE SIGNAL INTELLIGENCE

On the following pages, two of the most common steps that are done to signals when they are processed are covered. These are: *waveshaping*, or changing the shape of the signal; and *mixing* the signal with other signals or frequencies. In addition, the *frequency components* of certain types of signals are described in some detail, since they are the basis of signal shape both before and after distortion occurs.

mixing frequencies

In the processing of electronic signals, different frequencies are often *mixed* for various reasons. The products, or resulting signals, produced by mixing depend on whether the mixing is done in a *linear* or a *nonlinear* device. In electricity, a device, such as a resistor, is called linear because *equal changes* in applied *voltage* cause *equal changes* in *current.* Thus, the increase in current caused by an increase in applied voltage from 2 to 4 volts is the same as the current increase caused by a voltage increase of 100 to 102 volts. In other words, the *resistance* is essentially *constant* over the resistor's normal operating range.

A nonlinear device, on the other hand, has a *variable impedance,* whose value depends on circuit conditions. Thus, in a nonlinear device, a 2-volt increase in applied voltage from 2 to 4 volts may cause the current to increase 1 ampere; but the same 2-volt increase from 100 to 102 volts may cause the current to increase 2 amperes.

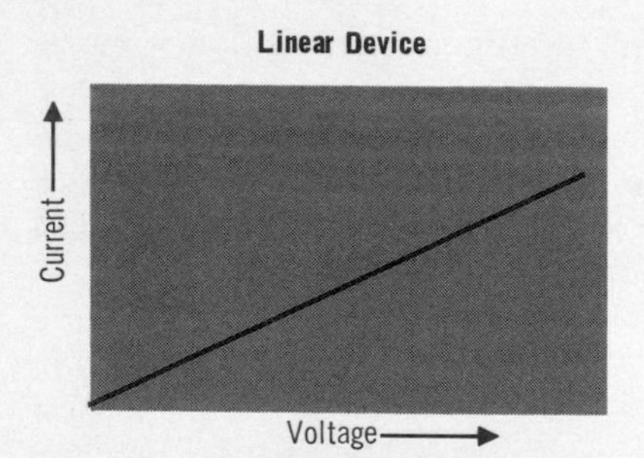

The voltage-current relationship in a linear device is represented by a straight line, since equal voltage changes produce equal current changes

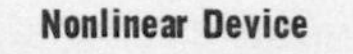

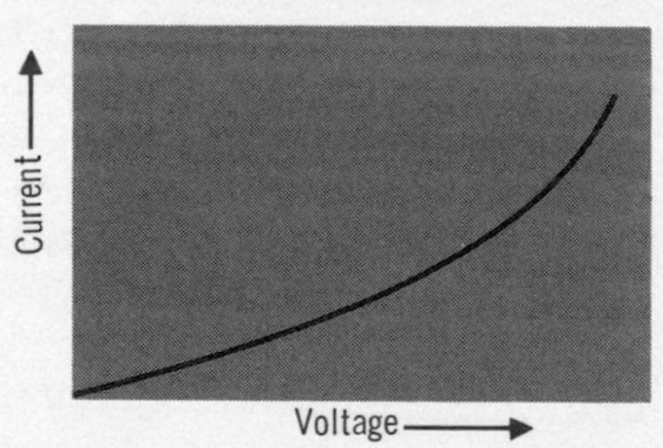

The voltage-current relationship in a nonlinear device is represented by a curved line, since equal voltage changes produce different current changes

When two waves are simultaneously applied to a linear device, they combine to form a new *complex* wave. This wave has a shape that is different than either of the two original waves, and contains as components the frequencies of the two waves. However, and this is the significant point, the only frequencies the new wave contains are those of the original waves. *No* new frequencies are produced. In nonlinear devices, as you will see, the situation is entirely different.

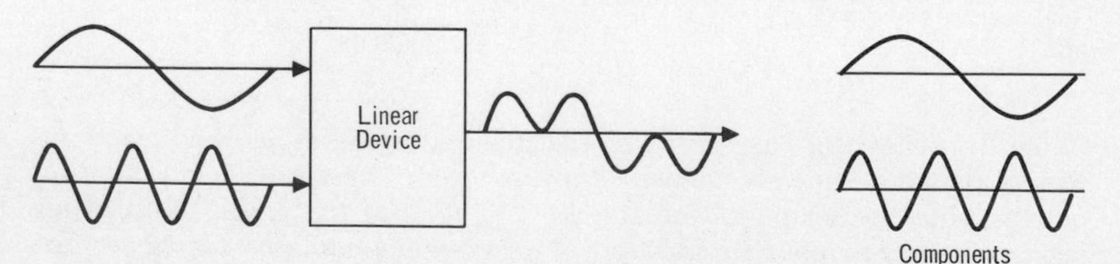

When two waves are mixed in a linear device, the resulting complex wave contains only the two original frequencies

nonlinear mixing

When two waves of different frequencies are mixed in a *nonlinear* device, the result is a complex wave, the same as in linear mixing. However, whereas in linear mixing the resultant wave contains only the frequencies of the original signals, in nonlinear mixing the resultant wave contains *additional* frequencies. These additional frequencies are produced by the mixing process. The total number of additional frequencies produced depends on the specific nonlinearity of the mixing device. Generally, however, devices are used with characteristics such that the resultant wave contains the two original frequencies, plus an additional frequency equal to their *sum* and another equal to their *difference*. Thus, if a 9-kHz and a 10-kHz wave are mixed in a nonlinear device, the resultant wave has frequency components of 9, 10, 19, and 1 kHz.

You probably recognize the similarity between *nonlinear mixing* and *modulation*. Actually, they are essentially the same process, with the difference between them being the frequencies of the input waves, and the components of the output wave that are used. In modulation, a high-frequency carrier is mixed with a relatively low-frequency modulating signal, and the carrier frequency and sum and difference frequencies are used. In what is called mixing, as you will see, two high-frequency waves are mixed, and generally only their *difference frequency* is used. This type of mixing is also called *heterodyning*.

Quite often, if the sum and difference frequencies are strong enough, they can mix to form more sum and difference frequencies that will beat with themselves and the original frequencies to form even new ones. However, the amplitudes of sum and difference frequencies are generally small, so continued heterodyning is usually negligible.

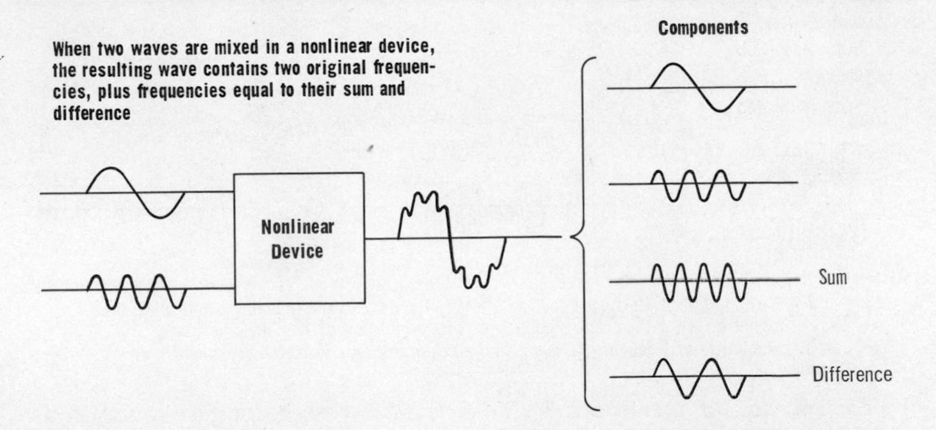

When two waves are mixed in a nonlinear device, the resulting wave contains two original frequencies, plus frequencies equal to their sum and difference

In actual circuits, more frequencies than these are usually produced by the mixing process. For most practical purposes, though, these other frequencies can be neglected, since the amplitudes of the sum and difference frequencies get smaller and smaller in any continued mixing process

heterodyning

In practically all types of electronic receiving equipment, received signals must be amplified, or built up, before they are demodulated. For reasons that you will learn later, it is highly desirable that this amplification be done at a *single* frequency rather than at the frequencies of the many different carriers that are received. To make this possible, all modulated carriers that are received are first converted to a common frequency, called the *intermediate frequency* (if). The intermediate frequency is produced by the nonlinear mixing of the received signal with a high-frequency sinusoidal wave, called the *oscillator frequency*, produced in the receiver. Either the sum or difference frequency produced by the mixing can then be used for the intermediate frequency. In most cases, it is the *difference frequency* that is used. All the other frequencies are removed by filtering.

For the intermediate frequency to be the same regardless of the carrier frequency of the received signal, the oscillator frequency must be changed each time a new carrier frequency is received. In most receivers, such as a common AM radio, for example, this is done by arranging the controls so that the oscillator frequency is changed by the required amount at the same time that a new carrier is tuned in.

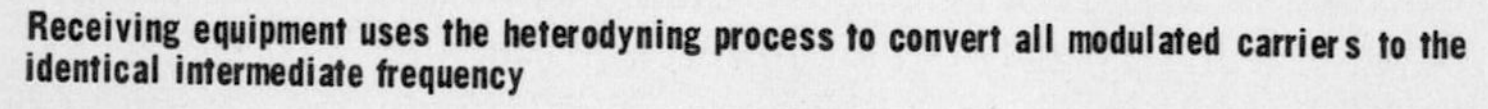
Receiving equipment uses the heterodyning process to convert all modulated carriers to the identical intermediate frequency

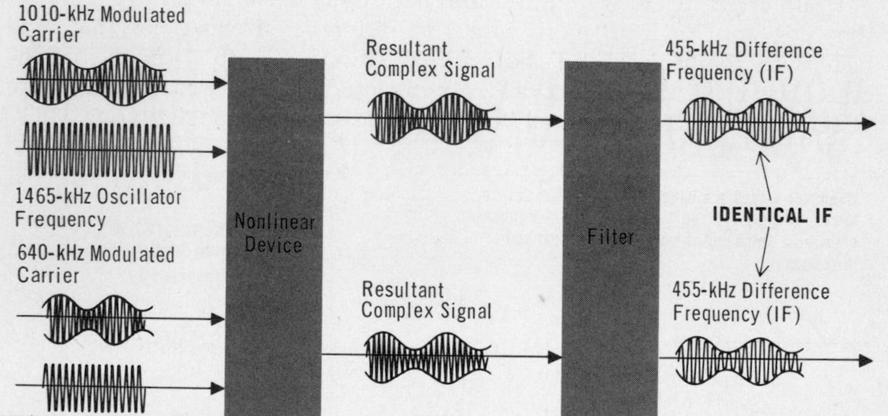

For a 455-kHz intermediate frequency, as shown, the oscillator frequency must always be 455 kHz higher or lower than the frequency of the modulated carrier. In practically all cases, the oscillator frequency is made higher than the carrier frequency

This mixing of the received signal with an oscillator frequency and then using the sum or difference frequency as the receiver intermediate frequency is called *heterodyning*. The signal intelligence is not affected by heterodyning. It is only put on a carrier of a different frequency. The difference frequency, normally used as the intermediate frequency, is also called the *beat frequency*.

producing the beat frequency

When two waves are heterodyned, they produce a resulting wave that contains the two original frequencies plus their sum and difference. The different, or *beat*, frequency is then selected by filtering and used as the intermediate frequency. As shown, when two waves are mixed, they *reinforce* each other at some points and *oppose* each other at others. This addition and subtraction of the instantaneous amplitudes produces a wave having a *varying amplitude*. The pattern, or *envelope*, of the varying amplitude is exactly equal to the difference between the frequencies of the original waves. Thus, as shown, if a 10-Hz wave is mixed with an 8-Hz wave, they produce a wave whose amplitude varies at a rate of 2 Hz, which is the beat frequency.

As was mentioned, the beat frequency produced by heterodyning a modulated carrier with an oscillator frequency contains all of the intelligence of the modulated carrier. Only the intelligence is carried on the *intermediate frequency* rather than the higher r-f carrier frequency. As you can see, the variations of *peak-to-peak amplitude* of the beat frequency follow the same pattern as those of the AM carrier. And since the carrier variations represent the intelligence, the intelligence has, in effect, been transferred to the beat frequency. In summary, then, heterodyning of an AM carrier causes the amplitude variations of the carrier to be transferred to a lower frequency wave.

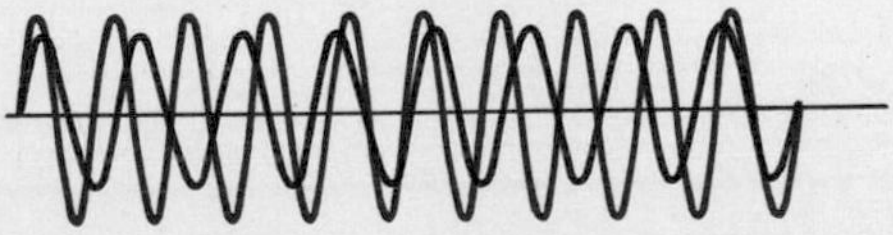

10-Hz Wave + 8-Hz Wave

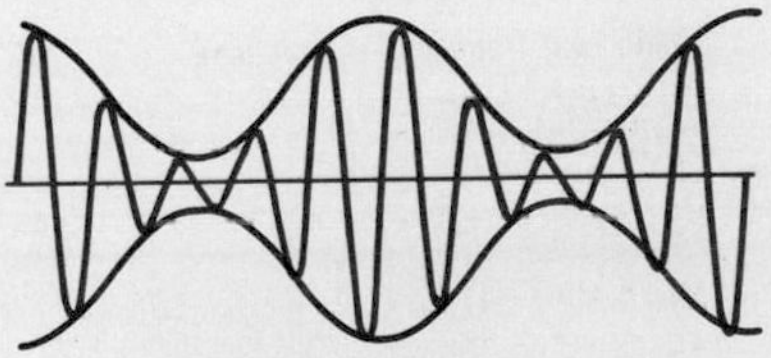

Wave Produced by Addition of Instantaneous Amplitudes

Envelope Corresponds to 2-Hz Wave

The beat, or difference, frequency is the envelope of the varying-amplitude wave produced by adding the instantaneous amplitudes of the two waves being mixed

Heterodyning of an FM carrier is similar, except that it is *frequency variations* that are transferred to a lower frequency wave. The beat frequency is a *band* having a width corresponding to the signal bandwidth, and a center frequency equal to the desired intermediate frequency.

Beat frequency detection is often used with interrupted CW carriers that do not have tones. The carrier is mixed with an oscillator signal that produces a difference frequency that can be heard. Then, as the carrier goes on and off, the audible difference frequency goes on and off.

the beat frequency and intelligence

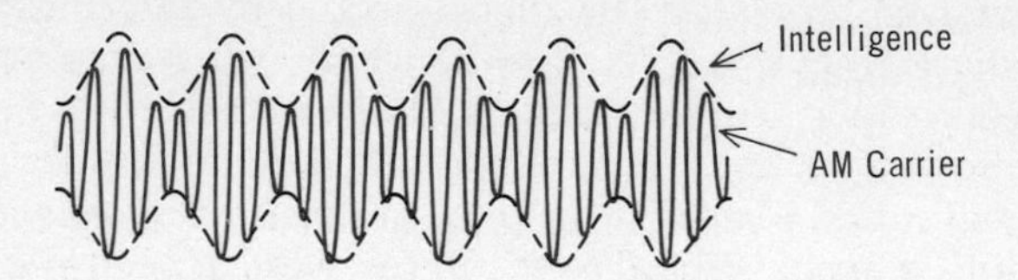

When this AM carrier...

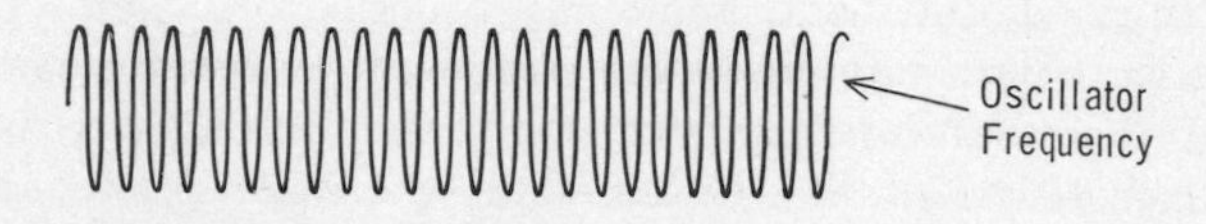

...Is heterodyned with this oscillator frequency...

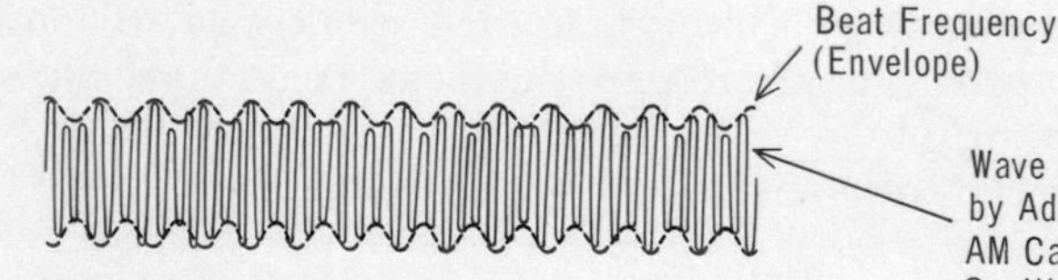

...This beat frequency is produced...

...But since the peak-to-peak amplitude of one of the heterodyning waves varied, so does the peak-to-peak amplitude of the beat frequency

The variations of the peak-to-peak amplitude of the beat frequency are the same as the variations of the peak-to-peak amplitude of the original AM carrier. And it is the variations that represent the intelligence

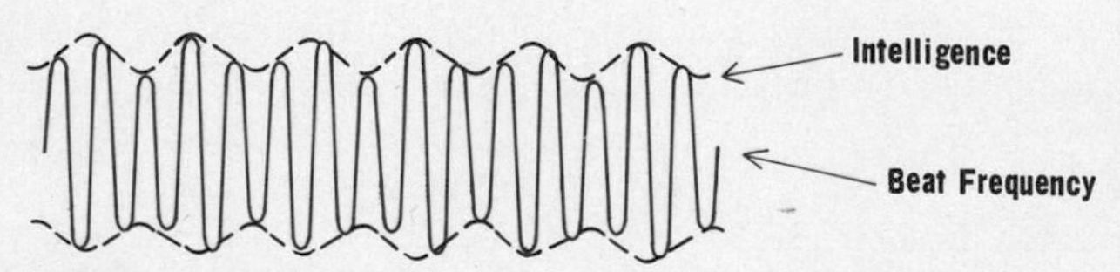

After removing the unwanted frequencies by filtering, the beat frequency alone carries the intelligence

summary

☐ Facsimile is a process that uses electronic signals to transmit visual material. The signals are generated photoelectrically, transmitted, and then converted to images on paper by one of several means. ☐ The Federal Communications Commission allocates frequency bands for all radio transmission within the United States. It also limits the amount of power that can be transmitted by various types of signals.

☐ A nonlinear device has a variable resistance, whose value depends on the applied voltage. ☐ A linear device has a linear resistance. That is, equal changes in applied voltage cause equal changes in current. ☐ When two signals are mixed in a nonlinear device, heterodyning takes place. ☐ A wave produced by heterodyning contains four components. These are the two original frequencies, plus their sum and difference frequencies. ☐ Modulation and heterodyning are essentially the same process. In modulation, one of the original frequencies (the carrier) plus the sum and difference frequencies are used. In heterodyning, only the difference frequency is usually used.

☐ Heterodyning is used in most receivers to convert all incoming modulated carriers to a single intermediate frequency. ☐ The intermediate frequency is produced by heterodyning the incoming signal with a local oscillator frequency, and using the resulting difference frequency. ☐ Signal intelligence is not affected by heterodyning. It is only put on a carrier of a different frequency. ☐ The difference frequency produced by heterodyning is often called the beat frequency. ☐ Beat-frequency detection is frequently used with interrupted CW carriers that do not have tones. This allows an audible difference frequency to be heard as the carrier goes on and off.

review questions

1. Briefly describe how a facsimile signal is developed.
2. The Standard Broadcast Band extends over what frequency range?
3. What characterizes a *nonlinear device*?
4. Is a resistor a linear device?
5. What is the difference between heterodyning and nonlinear mixing?
6. If a 5- and a 10-kHz wave are mixed in a nonlinear device, what are the frequency components of the resulting wave?
7. How does modulation differ from heterodyning?
8. What is an *intermediate frequency*? How is it produced?
9. What is a *beat frequency*?
10. Does heterodyning affect signal intelligence?

the basic signal and its harmonics

Voice signals contain *many different frequencies;* the combination of these frequencies gives the voice signal its shape. Each of these frequencies is a *sine wave,* but when they are combined, the resulting signal is not sinusoidal. This is true of every other signal that can be represented by a waveshape, except for a pure sine wave, which consists of only one frequency. All other waves consist of many frequencies having *different* amplitudes and phase relationships.

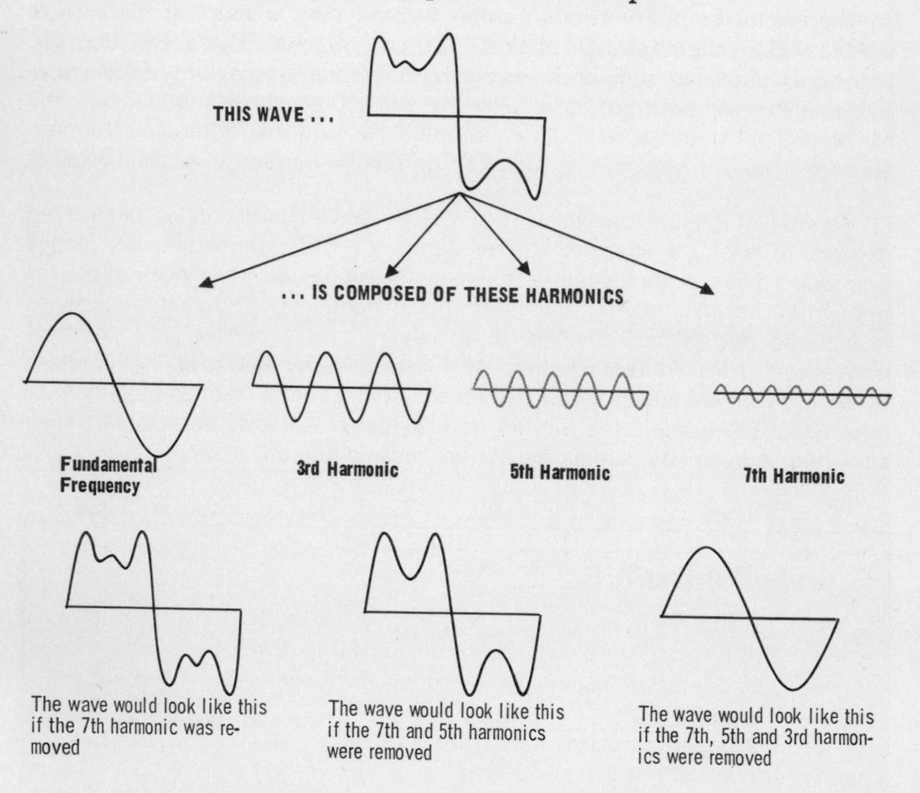

The *component frequencies* of any wave are all multiples, or *harmonics,* of some *basic frequency,* which is called the *fundamental frequency;* this is the lowest frequency in the wave. Thus, if a wave has a fundamental frequency of 100 Hz, its second harmonic is 200 Hz, its third harmonic 300 Hz, and so on. The first harmonic is actually the fundamental frequency.

All of the odd numbered harmonics of a wave are called the *odd harmonics;* and all the even numbered ones, the *even harmonics.* As you will see later, some waves contain only even harmonics, some contain only odd harmonics, and others contain both.

the harmonic content of pulses

Since the harmonics contained in a wave are responsible for its shape, anything that disturbs the harmonic content *distorts* the *waveshape*. If the shape of a signal is to be preserved exactly, therefore, the circuits that process the signal must *respond equally* to all of the frequencies contained in the signal. In practice this is impossible to do, since many circuit components, such as inductors and capacitors, respond differently to different frequencies. As a result, some distortion of a signal always takes place. This is not too bad, though, as long as the harmonics *most responsible* for the signal's shape are not affected by the circuit components. For some signals, only a few harmonics are necessary to satisfactorily preserve the waveshape. For other signals, especially those in which the waveshape changes rapidly, many more harmonics are required.

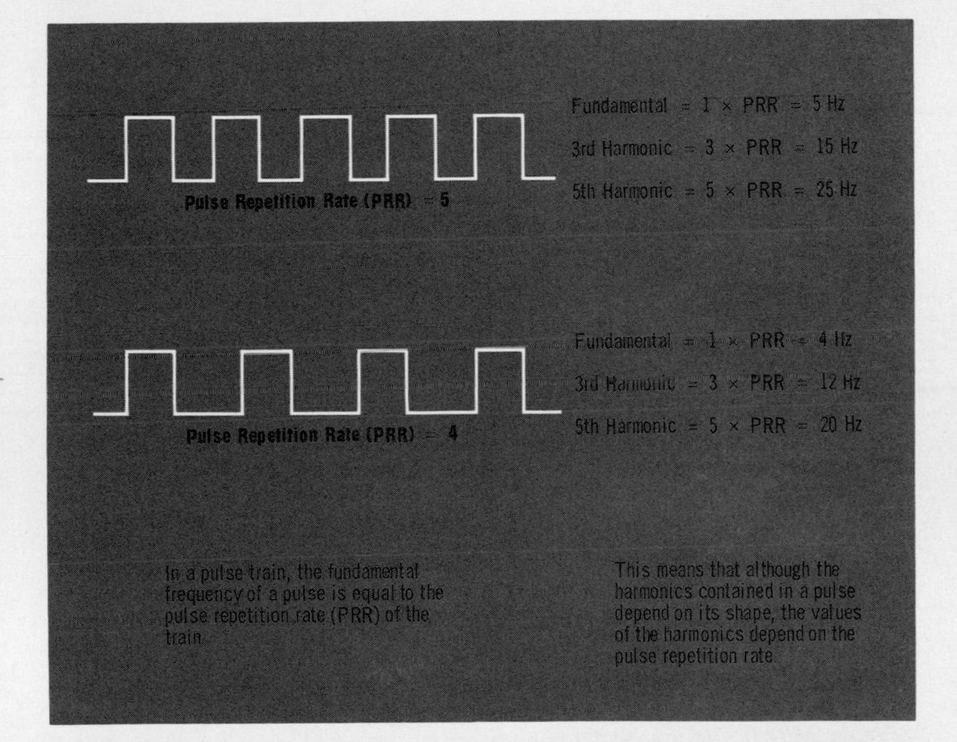

In a pulse train, the fundamental frequency of a pulse is equal to the pulse repetition rate (PRR) of the train

This means that although the harmonics contained in a pulse depend on its shape, the values of the harmonics depend on the pulse repetition rate

The harmonic content of pulses is important not only from the standpoint of preserving waveshape, but from the standpoint of *deliberately* changing waveshape as well. Very frequently in electronic circuits, it is desirable for one reason or another to change the shapes of pulses without destroying their basic pulse nature. Many of the circuits that perform this waveshaping effectively do so by changing the harmonic content, and therefore the shape, of the pulses.

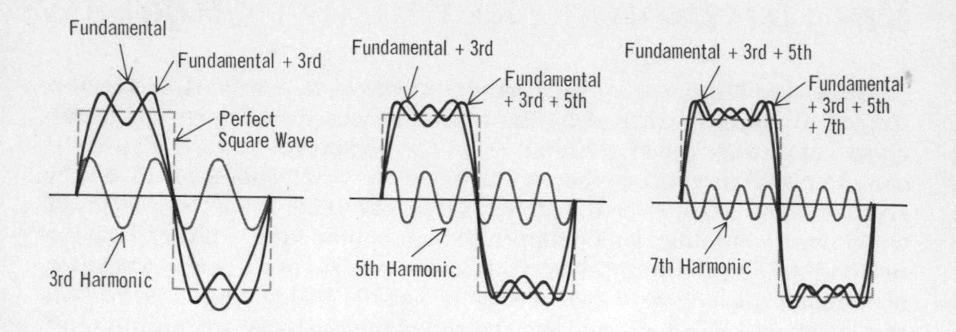

As more and more harmonics are added, a square wave comes closer to its theoretically perfect shape

the square wave

A perfect square wave has straight vertical sides, which indicates *zero* rise and decay times. Such a theoretically perfect wave has an *infinite* number of *odd* harmonics and *no even* harmonics. Since an infinite number of harmonics requires an *infinite bandwidth,* perfect square waves are impossible to achieve in electronic circuits. However, by using the fundamental and the lowest nine odd harmonics (3 through 19), practical square waves can be achieved that closely resemble the perfect square ware.

The effect of limiting the harmonics to a finite rather than an infinite number is to cause a *sloping* of the sides of the square wave. This means that the rise and decay of the pulse takes a certain amount of time, rather than occurring instantly. The more harmonics that are included, the steeper is the slope of the sides, and therefore the shorter are the rise and decay times.

In any pulse, the *higher harmonics* have the most effect on the *rise* and *decay* times of the pulse, while the *lower harmonics* have the most effect on the *duration* of the pulse. Thus, a circuit that does not respond to the higher frequencies will distort a square wave by increasing its rise and decay times. A circuit that does not respond to the lower frequencies, on the other hand, will distort the flat horizontal portions of a square wave.

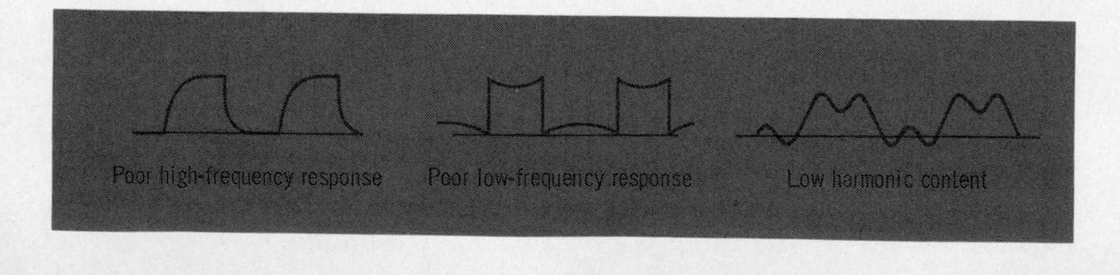

the sawtooth wave

The sawtooth wave is one of the most useful waveshapes in electronics. Its gradual rise and then abrupt drop to zero is the basis for many circuits that operate on a linear increase in voltage followed by an instantaneous drop in voltage to zero. A perfect sawtooth wave contains an infinite number of both odd and even harmonics. In other words, it contains all possible harmonics. As you know, in electronic circuits this requires an infinite bandwidth, so perfect sawtooth waves are never realized. Nevertheless, almost perfectly shaped sawtooth waves can be obtained by including a reasonable number of odd and even harmonics.

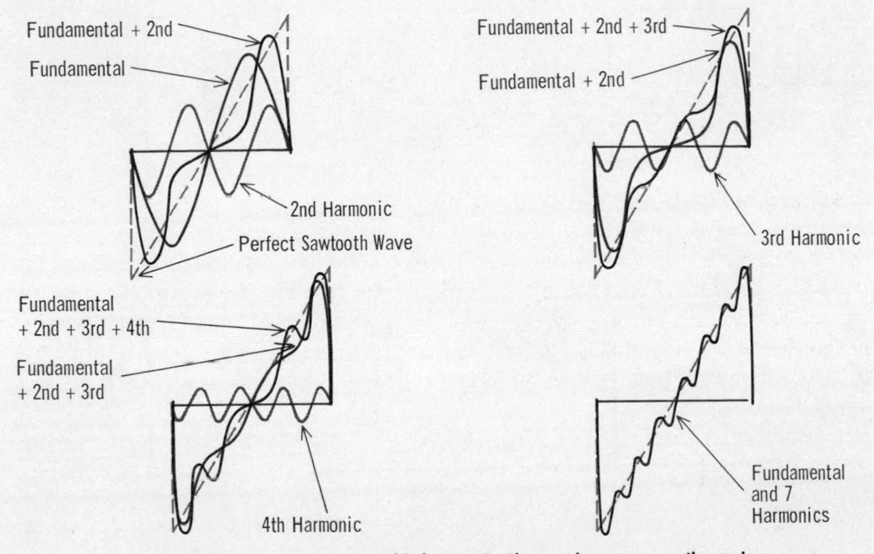

As more and more harmonics are added, a sawtooth wave becomes smoother and smoother, and comes closer to the shape of the perfect sawtooth

In a sawtooth pulse, the lower harmonics affect the rising portion of the pulse, while the higher harmonics affect the decay time. Thus, as shown, a circuit with poor low-frequency response will steepen the slope of the rising portion of the pulse. This has the effect of reducing the rise time. A circuit with poor high-frequency response will cause the pulse to decay more gradually. If the decay is slow enough, it can even run into the rise time of the next pulse.

Poor low-frequency response

Poor high-frequency response

Low harmonic content

waveshaping

A number of electronic circuits have the sole function of *changing* the *shape* of a signal or wave. This is especially true of pulse-type waves, although not limited exclusively to them. The particular way in which a circuit changes the shape of a wave depends on the characteristics of both the *wave* and the *circuit*. Because of this, not only will the same wave be shaped differently by different circuits, but the same circuit will have a different effect on various types of waves. As a result, waveshaping is a complex subject, requiring a knowledge not only of waves and their components, but of circuits and their properties as well.

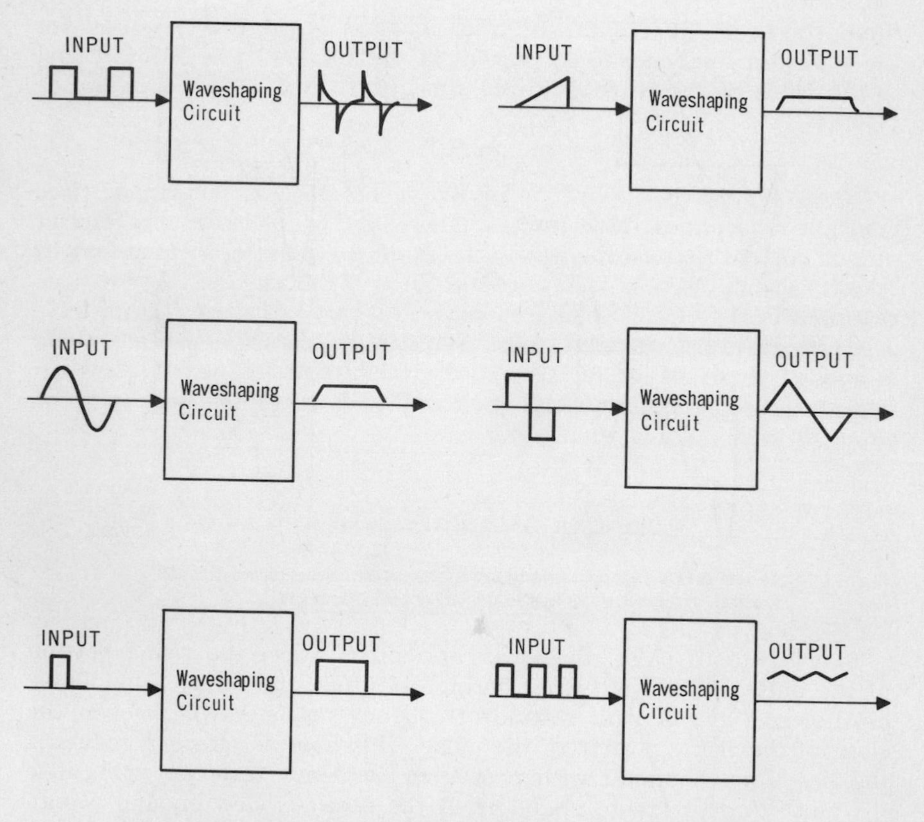

In practically all electronic equipment, waveshaping of some sort takes place. Sometimes the shape of the wave is changed only slightly. In other cases, though, a wave may be changed to a completely different shape

Nevertheless, there are certain types of waveshaping operations that are used extensively, and that involve the same basic principles regardless of the types of waves they are reshaping. Before these operations are described, though, a brief review of RL and RC *time constants* will be given.

the time constant

You should recall that the time constant of an RL or RC circuit is a measure of how *rapidly* the voltages and currents in the circuit can respond to changes in input voltage. A *small* time constant indicates that the circuit can respond *very rapidly*. A *large* time constant indicates that the circuit responds *very slowly*. Mathematically, the time constant of an RL circuit is the time required for the current to build up to 63 *percent* of its *maximum* value. It is equal to the value of inductance (L) divided by the value of resistance (R).

$$t = L/R$$

Similarly, in an RC circuit, the time constant is the time required for the capacitor to charge to 63 percent of the applied voltage. It is equal to the value of the resistance (R) times the value of the capacitance (C), or

$$t = RC$$

When a pulse is applied to an RL or RC circuit, the circuit time constant determines three factors. These are (1) *whether* the circuit output voltage rises to the peak voltage of the pulse; (2) *how long* it takes to rise to the peak voltage of the pulse, if it does; and (3) the time required for the circuit voltage to *decay*. All three of these factors have a significant effect on pulse shape. You can see then that when a pulse is applied to an RL or RC circuit, the circuit time constant plays an important role in determining whether the shape of the pulse will be changed, and, if so, in what way.

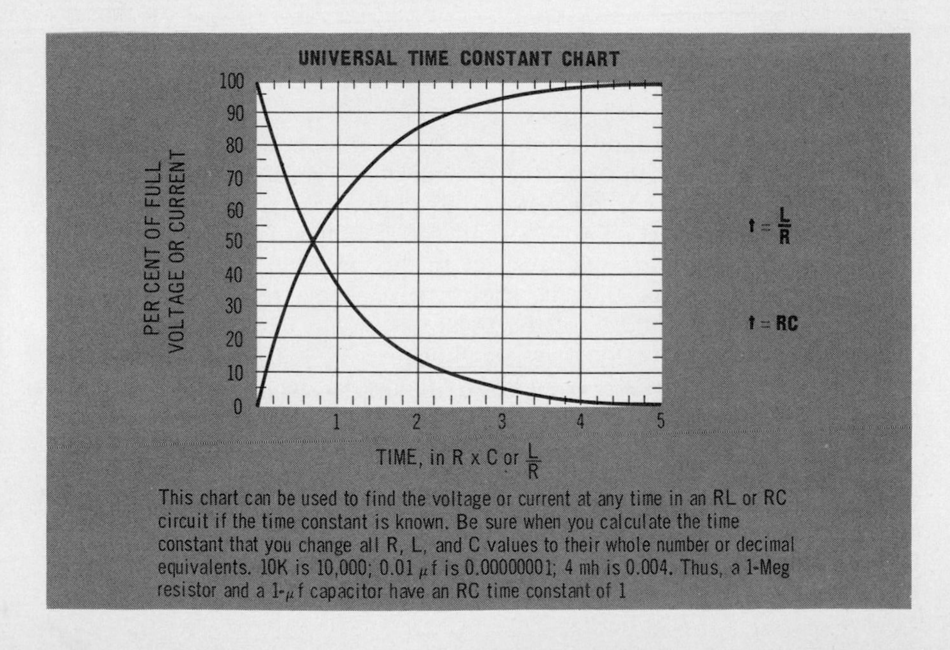

This chart can be used to find the voltage or current at any time in an RL or RC circuit if the time constant is known. Be sure when you calculate the time constant that you change all R, L, and C values to their whole number or decimal equivalents. 10K is 10,000; 0.01 μf is 0.00000001; 4 mh is 0.004. Thus, a 1-Meg resistor and a 1-μf capacitor have an RC time constant of 1

time constant and signals

In itself, a time constant is neither *long* nor *short*. It is only when the time constant is referred to some *reference period* that the terms "long" and "short" become meaningful. With a *sine wave*, the time constant is compared to the time *period* of one cycle. In the case of a pulse, other reference periods also of interest are usually the *rise time*, the *duration*, and the *decay time*. With respect to these reference periods, plus the cycle period, a time constant can be considered as being long in one case, and short in another.

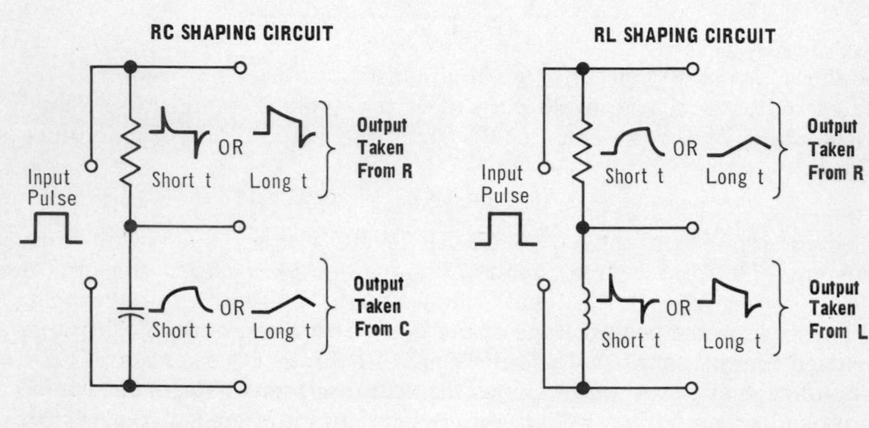

The identical input pulse is shaped differently, depending on whether the time constant is long or short with respect to the pulse duration. The resulting pulse shape depends also on whether the output pulse is taken from the resistor or the reactive component of the circuit

The pulse taken from the resistor of the RL circuit is identical to that taken from the capacitor of the RC circuit. Similarly, the pulse taken from the resistor of the RC circuit is identical to that taken from the inductance of the RL circuit

A commonly used rule of thumb is that a time constant is long if it is over *five times* the duration of the reference period. Similarly, if a time constant is less than *one-fifth* the duration of the reference period, it can be considered as short. You can see that it is very possible for a time constant to be long with respect to the rise and decay times of a pulse, and at the same time be short with respect to the duration, or constant-amplitude portion, of the pulse.

Knowing the relative longness or shortness of a time constant with respect to a pulse is still not sufficient to determine the effect that an RL or RC circuit will have on the pulse. This is because the output waveform of such a circuit depends on whether the output is taken from across the *resistor*, or from across the *reactive component* (inductor or capacitor). Completely different waveforms can be produced, depending from where the output pulse is taken.

With sine waves, the time constant determines the phase *shift* of the signal.

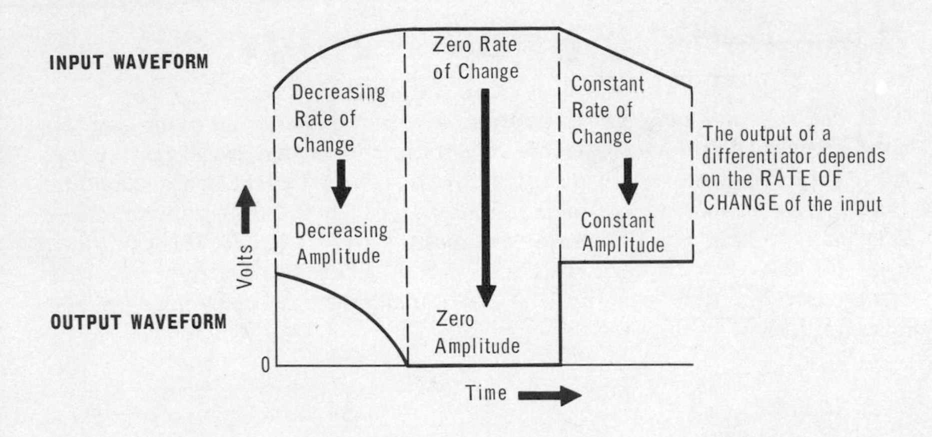

This output waveform would be produced by a perfect differentiator having an infinitely small time constant. Practical differentiated waveforms are shown below

the differentiator

A differentiator is a shaping circuit with a short time constant whose output voltage is proportional to the *rate of change* of the input voltage. When the rate of change is zero, the output is zero; when the rate is constant, the output is constant; and when it is increasing or decreasing, the output increases or decreases. Thus, it is the *changes* in the value of the input that produce the output, and not the values themselves.

Any *horizontal* line means a *zero* rate of change, and therefore zero output. Any straight line that is *not horizontal* means a *constant* rate of change, and so a constant output. The closer the line is to being perfectly vertical, the greater is the rate of change. A curved line represents an increasing or decreasing rate of change, and therefore a changing output. These relationships between input and output waveshapes are shown.

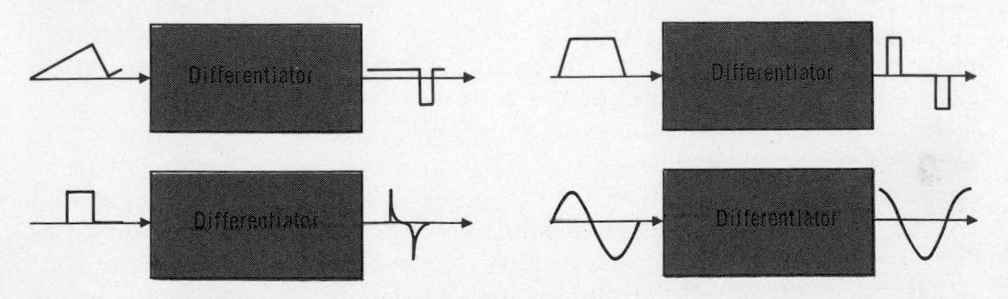

Notice that a differentiator does not change the shape of a sine wave. All it does is to shift the phase of the sine wave

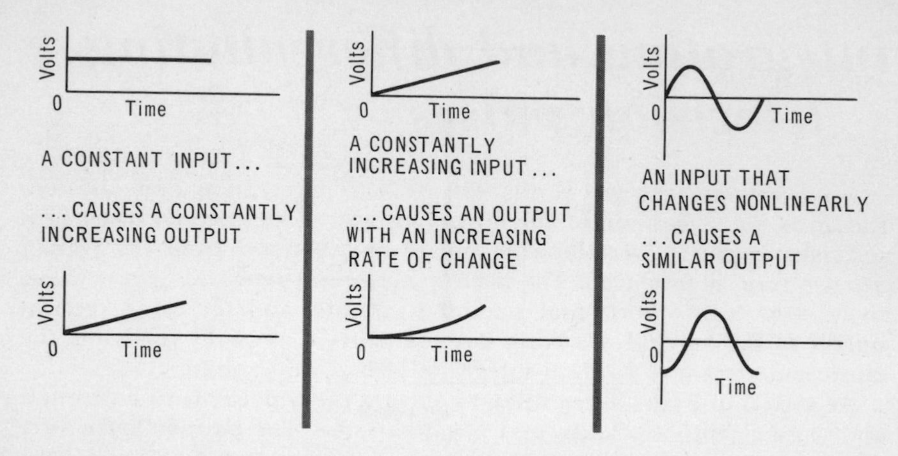

In an integrator, the rate of change of the output varies in accordance with the value of the input

the integrator

Another frequently used waveshaping circuit is the *integrator*. In many ways, the integrator functions exactly *opposite* to the differentiator. Whereas the *differentiator* is a *short-time-constant* circuit, the *integrator* is a *long-time-constant* circuit. In addition, whereas the output of a differentiator varies in accordance with the rate of change of the input, in an integrator, the rate of change of the output varies in accordance with the *value* of the input. Thus, if the input voltage to an integrator is constant, the output *increases* at a constant rate; while if the input increases or decreases at a constant rate, the output increases or decreases at an *increasing* rate. Also, if the input increases or decreases in a nonlinear way, the rate of change of the output varies accordingly. These input–output relationships are shown.

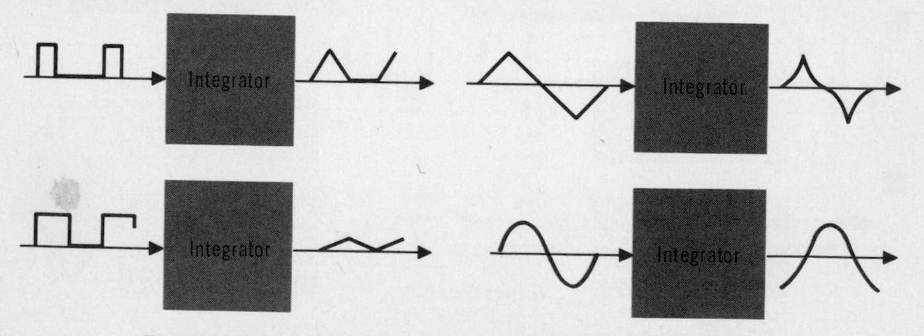

These are some of the changes in waveform shape that can be made by an integrator. Notice that the integrator, like the differentiator, shifts the phase of a sine wave, but does not change its shape

integrating and differentiating television pulses

An excellent example of the use of both integration and differentiation is the separation of the horizontal and vertical sync pulses of a television signal. After the sync pulses are removed from the signal, a pulse train is produced. The narrow horizontal sync pulses now have to be sent to the horizontal scanning circuits, and the wide vertical pulses to the vertical scanning circuits. This is done by applying the entire pulse train to both an integrator and a differentiator.

As shown in B, the differentiator converts *every* pulse, both horizontal and vertical, into a positive and negative *spike*. The positive spikes are all *equally spaced,* since they correspond to the *leading edge* of the pulses, which are all the same distance apart. The negative spikes are not equally spaced because of the difference in widths between the horizontal and vertical pulses. This is unimportant, though, since all of the negative spikes are removed before the spike train is sent to the horizontal circuits. Thus, the differentiator provides a pulse train of equally spaced spikes for horizontal synchronization.

At the same time that the differentiator is converting the sync pulses into spikes, the integrator is integrating each and every sync pulse. As shown in C, integration of the horizontal sync pulses produces low-amplitude sawtooth pulses that start at the zero reference and *end* there also. But when the vertical sync pulses are integrated, their greater width produces a somewhat triangular wave that does *not* return to the zero reference. Succeeding pulses, therefore, have *greater amplitudes,* and in this way, a single, high-amplitude vertical pulse is built.

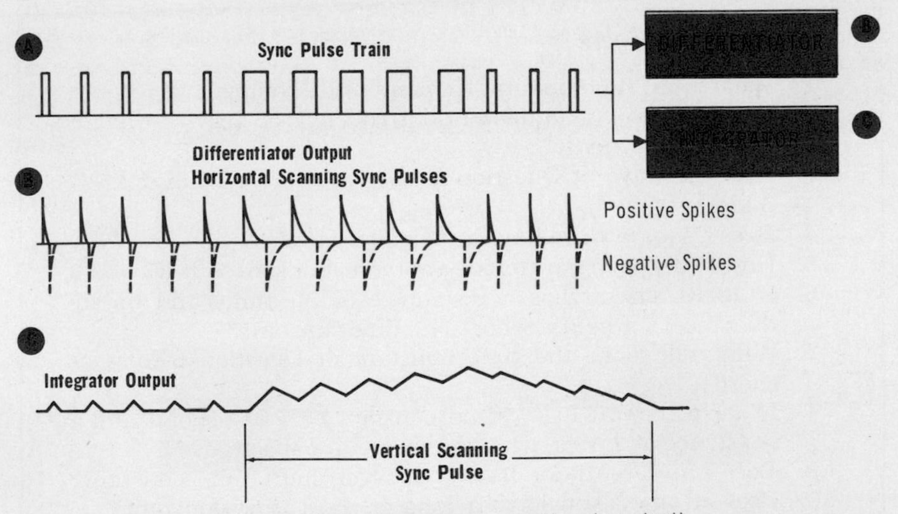

The pulse train shown above is actually only a simplified version of the vertical television pulse train

summary

□ All complex waves consist of many sine-wave components having different amplitudes and phases. □ The lowest component frequency in a complex wave is called the fundamental frequency. All other component frequencies are harmonics of the fundamental frequency. □ All of the odd numbered harmonics (1, 3, 5, etc.) of a wave are called the odd harmonics. All of the even numbered harmonics (2, 4, 6, etc.) are called the even harmonics. □ The shape of a wave is determined by the harmonics it contains. Changing the harmonics, therefore, changes the waveshape. □ A perfect square wave has an infinite number of odd harmonics and no even harmonics. A perfect sawtooth wave contains an infinite number of both odd and even harmonics. These waves require an infinite bandwidth, and so are impossible to achieve exactly.

□ The time constant of an RL circuit is given by $t = L/R$, and that of an RC circuit by $t = RC$. □ A small time constant indicates that the circuit can respond to an input signal very rapidly. A large time constant indicates the circuit responds very slowly. □ The time constant of a circuit determines whether the shape of an applied pulse will be changed by the circuit. □ Generally, a time constant is considered long if it is over five times longer than the duration of the reference period being used. It is short if it is less than one-fifth the duration of the reference period.

□ A differentiator is a short-time-constant circuit whose output is proportional to the rate of change of the input voltage. □ An integrator is a long-time-constant circuit in which the rate of change of the output varies in accordance with the value of the input.

review questions

1. What is the fundamental frequency of a complex wave?
2. If a wave has a fundamental frequency of 2 kHz, what is the third harmonic?
3. Does the wave of Question 2 have a component of 1 kHz? 3 kHz?
4. Does a square wave have any odd harmonics?
5. How many even harmonics are there in a sawtooth wave?
6. If an RL circuit has a resistance of 50 ohms and an inductance of 1 henry, what is its time constant?
7. What will make the time constant of Question 6 long or short?
8. If an RC circuit has a time constant of 0.01 second and a resistance of 1 Meg, what is the circuit capacitance?
9. Does a differentiator have a long or short time constant?
10. Does an integrator have a long or short time constant?

index

index